SpringerBriefs in Psychology

For further volumes:
http://www.springer.com/series/10143

Henry Kellerman

The Discovery of God

A Psychoevolutionary Perspective

 Springer

Henry Kellerman
Postgraduate Psychoanalytic Society and Private Practice
New York, NY, USA

ISBN 978-1-4614-4363-6   ISBN 978-1-4614-4364-3 (eBook)
DOI 10.1007/978-1-4614-4364-3
Springer New York Heidelberg Dordrecht London

Library of Congress Control Number: 2012940748

Printed on acid-free paper

Springer is part of Springer Science+Business Media (www.springer.com)

Books by the Author

Coauthored Books
(with Anthony Burry, Ph.D.)

Psychopathology and Differential Diagnosis: A Primer

Volume 1. History of Psychopathology

Volume 2. Diagnostic Primer

Handbook of Psychodiagnostic Testing: Analysis of Personality in the

Psychological Report. 1st edition, 1981; 2nd edition, 1991;

3rd edition, 1997; 4th edition, 2007.

(Japanese edition, 2011).

Edited Books

Group Cohesion: Theoretical and Clinical Perspectives

The Nightmare: Psychological and Biological Foundations

Coedited Books
(with Robert Plutchik, Ph.D.)

Emotion: Theory, Research, and Experience

Volume 1. Theories of Emotion

Volume 2. Emotions in Early Development

Volume 3. Biological Foundations of Emotion

Volume 4. The Measurement of Emotion

Volume 5. Emotion, Psychopathology, and Psychotherapy

In memory of

Henry Bender, Ph.D.

and,

The Honorable Bernard Becker,

Bernie and Hank

Devoted forever-friends

Contents

Introduction

At the beginning of this second decade of the twenty-first century, the controversy regarding a belief in the presence of God (or, in the presence of a God) has heated up. A polarized discourse, it essentially arranges theists against atheists. In broad terms, theists view atheists as adversarial, while atheists see theists as smugly ensconced in their traditional acceptance of faith in God. Thus, atheists proclaim that theists smear them with such terms as "subversive," "obscene," "blasphemous," and "sacrilegious," descriptions that denigrate and lead to ostracism. Either way this search to resolve the issue of God's existence can be formulated as axiomatic (a self-evident truth) for believers because the believer assumes it is true. However, for nonbelievers (atheists), or even for doubters (agnostics), any automatic belief in God's existence is considered an uncontested assumption. Uncontested assumptions can certainly be incorrect, and if proved incorrect they then become successfully contested.

Meanwhile, in the absence of such proof, we witness the persistence of such questions as the following: When did man find God; when did man find the idea of God; how did man find God; why did man find God; why did man find the idea of God; or why and when did God find man.

The Issue of Origin

The evolutionary God (the origin, or discovery of God) could not be based simply on a consideration of biological variables such as the appearance of a cerebral cortex/thinking brain. Rather, in other terms the "discovery" of God from a psychoevolutionary perspective would need to include the concept of cognition. Without man's primary capacity to think (and to feel emotion entwined with thinking), the conceptualization of an extant God would not be possible. How does one think of God if one cannot think? Yet, of course, this does not rule out the existence of God.

This means that in so-called lower forms of life (in animals—from the ape down to the amoeba), the issue of the worship of God and the rituals associated with such worship do not exist. And even though feeling and thinking, or what might be considered proto-thinking, do exist (especially among higher-order phylogenetic species), nevertheless advanced complex thinking and cognitive skills are not present.

And this brings us to the thinking brain and the human capacity for complex conceptualization and emotional experience. The image of a God is necessarily connected to this thinking brain and to this feeling human person even though a believer will not want to acknowledge that the human mind may have created the image of God. But even if man was created in the image of God, then an implicit as well as an explicit connection exists between the two, God and man. Depending on which came first, God or man, the vestige of God would be in man, and/or the vestige of man would be in God.

And that is what this book is about. Does it all depend on God or does it all depend on us? We are concerned here not only with the issue of absolutes (there is a God/no God exists). We are concerned also with how man thinks about God, and that because man can think, then God becomes a specific thinking possibility.

And so the title of this book is: *The Discovery of God: A Psychoevolutionary Perspective.* If this book were to have been materialized by a turtle or a giraffe, the title could not possibly have been the same. If materialized by a turtle or a giraffe, the title would necessarily be one that subtracts the adjective "Psycho." The noun "psycho" implies a sophisticated convergence and fusion of thinking and feeling in a highly complex bio/neuro/cognitive organization of "psyche," only presumably relevant, relegated and referred by any scientific test, to humans.

Thus, to investigate the psychoevolutionary perspective applied to the phenomenon described as God, it is necessary to begin by trying to understand the issue from the vantage point of the human mind as well as how individuals and groups treat religion, and to follow the trail of the vast phylogenetic evolutionary process leading in contemporary biological/archeological history to the thinking brain.

What It Is All About

In a purely historical analysis, man's gradual acknowledgment or discovery of God can be traced from the dawn of hominid evolution and human culture into the era of Homo sapiens development.

The added dimension hypothesized here and considered to be the salient factor in man's discovery of God is not the socio/anthro/psycho/spiritual analysis that has been often proposed. The key here is that man's capacity to think about God, or to encounter God, includes an anatomical consideration; that is, biological evolution is certainly in its core meaning a tale of anatomical transformations. As stated above, ultimately the perception of God required an anatomical brain that could think complex thoughts. And then, in considering the gradual appearance in evolution of the cerebral cortex or thinking brain, the question may be posed as to what it was in previous anatomical structures that may have been a precursor or antecedent to this discovery that had a kind of homologous relationship (relationship of function) to the human brain. Such an analogous relationship between anatomical structures (and functions) of separate evolutionary eras may be appreciated especially with respect to "function"—to resemble one another. In other words, in an earlier evolutionary era, the function of some anatomical organ could be transferred or transmuted to a later evolutionary appearance of another organ despite the vestigial presence of the antecedant organ. Simply stated this is considered to be evolutionary development of "function" that becomes located in newer organs during such ongoing evolutionary development.

This homologous analogous-in-function structure as it relates to the location of God may have been the "tail" of animals that in evolution became vestigial. The tail of animals provides or provided protection from behind to assure greater physical security, to enhance the animal's gyroscopic sense of balance, to facilitate locomotion, and to even act as part of both the animal's defensive and offensive arsenal. All in all, the function of the tail that provided the animal with greater security, also, we might say, provided a corresponding homeostasis (peace of mind). Such function of the animal's tail constitutes the essence of the homologous anatomical structural analogy to the human thinking brain; that is, with respect to function, as the tail did for more primitive organisms, the thinking brain does for man. This brain provides man with the same ability to create methods of gaining greater physical security, ensuring a better emotional balance in the person, facilitating a better mastery of the environment, and generally also creating a defensive and offensive arsenal for protection—creating in essence, at the least, a chance for man to have relative peace of mind. Still further, it will be suggested that such peace of mind along with physical security and better emotional balance can be related to another property wired into DNA—that of the need as well as the expectation of fair play and a hope for empowerment along with the neutralization of any possibility of a dreaded disempowerment.

To see it this way is possibly to also see that it is not that religion was highly selected in evolution and then ultimately wired into our DNA. Rather, it was the need for protection—the particular *need for protection*—that was highly selected. So in this volume, *The Discovery of God: A Psychoevolutionary Perspective*, it is the combination of the disappearance of the tail in evolution (with respect to its function), along with the eventual appearance of a thinking brain (accompanied by an unconscious template of the need for the everlasting good parent) that will be the basis of our discussion, and from which this book has gained its title. By no means, however, is the evolutionary quest for protection as well as a need for the good parent in any way suggesting evidence regarding the veracity of either position—God/no God.

New York, NY, USA   Henry Kellerman

Chapter 1
God and the Important Anatomical Vestigial Structure

What was once a biological structure and eventually through evolution ultimately became vestigial, also became transformed or even transmogrified into the human psyche. It was not that this anatomical vestige became the brain. Rather, it was that certain of its functions were then assumed by the brain. From such transmutation of function or transmogrification, it is proposed that man was then able to identify God.

Though the ability to identify God may seem to imply the existence of a God, it is not my intention here to draw that implication. I am only referring to man's ability to posit a God. Such a proposition does not necessarily validate a God's existence nor does it refute such an existence.

Therefore, with respect to the discovery of God in terms of a psychoevolutionary perspective, I am here making reference to something in the human psyche that has become a surrogate anatomical vestigial evolutionary structure mediating a person's need for, and identification of God. It is proposed here that this anatomical vestigial structure was the "tail."

At this early point of the discussion regarding the discovery of God from a psychoevolutionary perspective, the hypothesis presented here considers the commonality between the function of the anatomical tail that became vestigial, and the newer cerebral cortex—the thinking brain with a prefrontal cortex development correlated to self-awareness. In this sense it is possible to understand that in evolution the tail disappeared in animals, and in function, resonated as a surrogate anatomical body part then appearing in an organism with an advanced cerebral cortex designed for understanding and "seeing" things—the thinking human brain—with a beginning neo-cortex development dated from approximately 150 million years ago. Essentially, psychoevolution (or evolutionary psychology) is the science of investigating and understanding the infrastructure of "mind." And of course included in the discussion of the mechanisms and vicissitudes of mind is the thematic strand marching through all of history—that of an inquiry into the issue of God. This psychoevolutionary position is also hinted at by Henig (2007) who asks (perhaps in a non-pejorative sense), whether religion is a vestigial artifact of a primitive mind. It is Henig's hint

H. Kellerman, *The Discovery of God: A Psychoevolutionary Perspective*,
SpringerBriefs in Psychology, DOI 10.1007/978-1-4614-4364-3_1, © The Author 2013

at the importance of tracing the roots of religion from such a psychoevolutionary vantage point.

But first, before entering directly into this discussion, we need to examine an analogy that perhaps can offer a useful set of tools to contribute to an understanding of the overall array of theories and thinking-postures regarding the idea, and even ideology of God. In this respect the psychology of empathy and gratitude needs to be examined. Essentially, this idea of empathy and gratitude in the context of a person's faith concerns the psychology of rapport. Therefore, in order to further examine theories on the ideology of God, an examination of the psychology of empathy as a key element of rapport can be useful as an approach to understanding the array of theories regarding the man/God issue.

Empathy and Gratitude

Before it is suggested as to how the appearance and experience of empathy relates to the entire context of God, and of worship of God, I believe it would be useful to examine the role of empathy in the psychotherapeutic endeavor with its implied optimism (and even hope), where one person seeks another in order to discuss and improve some aspect of life. Discussions of empathy are voluminously present in the psychological literature and can be sampled by the work of Blechner (1988), Josephs (1988), Kohut (1984), Levenson (1972), Mitchell (1988), and Rogers (1951).

For the most part, psychotherapists attach major importance to the phenomenon of empathy as it relates to the relationship between patient and therapist, especially with respect to the importance of establishing rapport. Rapport is a relationship benchmark largely characterized by a successful empathetic connection between the therapist and the patient. More specifically, rapport is more efficiently established when the therapist is able to accurately reflect the patient's feelings.

Empathy is not sympathy. It is not feeling sorry for the other person. It is the reflective expression of appreciation of the other and therefore represents an understanding of the other. This appreciation of the other person is validated by a sense that the therapist respects the patient's feelings. Thus, it is the therapist who is expressing the empathy. Correspondingly it is the empathetic therapist that is able to imagine what the patient feels by almost trying to be that person, that patient—to be able to put oneself, as it were, in the patient's shoes or the patient's skin; it is the ability to imagine oneself into the subjective experience of the other person, and therefore to know to a reasonably good extent, what that person feels.

Once the patient experiences the therapist's empathy, the patient will then feel not only understood, but also safe. The psychology of this phenomenon is expressed in the dictum: *Love is not* enough; that is, in order for a relationship to be nourished, each partner needs to convey that they appreciate and sufficiently understand the other—and importantly, to feel safe with the other. To convey understanding by reflecting the other's feeling is the actual underpinning to the empathy.

To feel understood is of course ubiquitously appreciated. Similarly, in prayerful homage, the person's relationship with God is almost isomorphic (exact in scale) with the same kind of search for understanding that any person seeks from a relationship partner.

The question becomes: in the psychotherapy session, does the successful achievement of empathy on the part of the therapist easily and even profoundly contribute to the aims of the therapy? The answer seems to be an obvious yes. Because the patient feels understood as a result of the therapist's empathetic presence (both in a general way and in what is specifically at hand—and said), then in fact, usually the patient will want to continue to work in the sessions. It can be reasonably proposed that the expression of empathy on the part of the therapist has an enormous positive effect; that is, to feel understood is powerful and therefore also contains an urge for more—for continued, further and sustained understanding.

Another important question concerns the possible negative effect to the empathy experienced by the patient. Therefore, is it possible there could be a downside to empathetic responses by the therapist to the patient—a negative effect to the patient feeling understood?

This question as to the possible or even hypothetical negative effect of empathy within the therapeutic endeavor is not often posed as a theoretical consideration in the scholarly approaches of clinical or scientific inquiry. Rather, it is generally accepted that empathy is a ubiquitous positive characteristic necessary for the momentum and work in the therapeutic process. Because of this assumed attitude about empathy, it may just be that something very important is absent in understanding a possible deeper aspect to the entire context and even infrastructure of empathy.

The proposed answer to the question as to whether empathy in the psychotherapeutic context—therapist to patient—can generate a negative effect in the therapy process is, I believe, yes. There is definitely a downside—even perhaps a danger at least to inordinate or excessive expressions of empathy—especially in the absence of any necessary confrontational encounters that challenge the patient.

And as a reminder, we are examining the issue of empathy as a prelude to developing an analysis of the discovery of God from a psychoevolutionary perspective; that is, from the vantage point of examining psychoevolutionary mechanisms permitting man to invent, find, or encounter God.

Empathy Has an Up/Down Personality

Initially, it should be stated that the therapist's emotional understanding or empathy toward the patient is not a panacea. In psychodynamically oriented psychotherapy (the attempt to identify cause and effect by examining the patient's history and also by seeing how this history impacts the person's present relationships as well as the therapeutic session itself), the absolutely strongest force of any of the vectors in such a therapeutic context is the patient's pathology (conflicts and disturbance).

And the main point to emphasize is that the patient's pathology will not in the slightest be affected by any sole empathy on the part of the therapist, except in cases where a so-called flight-into-health can be temporarily achieved by the patient's relief of tension based on such empathy. This is important to note because affecting pathology requires at least a mandatory raising of the patient's consciousness (the lifting of repression, or crystallizing a synthesis), and in most cases this raising of consciousness needs to contain subject matter of the patient's "dissatisfaction" or "anger" toward some other specific person in order for anything ultimately resembling progress to occur.

The flight-into-health is frequently illusional because of its usual temporary state. This sort of relief of tension is similar to the relief genuinely experienced by prayerful people who are reverential believers in the God that understands. Yet any cause and effect attribution between the psychology of empathy on the one hand, and evidence of the existence of God on the other, is purely on the basis of experiencing such empathy, actually illogical—not one logical iota above zero. In other words, the experience of empathy does not at all validate the actual existence of a God even though the prayerful person may experience it that way. On the other hand, even if the person experiencing empathy (and attributes it to God's understanding), and even if belief might be a function of a person's high suggestibility index, nevertheless such possible self-delusion does not at all necessarily invalidate the actual existence of a God.

After all is said and done, in the experience of acute tension and emotional pain, to receive empathy or to feel it emanating from one's presumed relationship to a God, is for the person declaring it, a huge dividend. In this sense, the alleviation of pain is, of course, very important.

The Negative Effect of Empathy: The Downside

People can experience all sorts of pathological variants, but anger that is buried in the unconscious, is usually the culprit in all of psychological pathology (Kellerman, 2008, 2009a). In order to get to this pathology, the objective in the therapy is to reduce the patient's psychological resistance to change thereby lifting some of the patient's presumed repressed material. Thus, the therapeutic process usually boils down to the attempt to identify hidden dissatisfaction (anger) toward a specific historical person in the patient's life that becomes displaced onto others (including the therapist) in the patient's current life.

In contrast, rather than facilitating this therapeutic goal of getting to the pathology, of reaching it, in each case of the expression of empathy, the empathetic response will in all likelihood contribute to even increasing the patient's resistance to this consciousness raising (regarding repressed anger). This is an important point because in order to gain access to personal emotional/psychological material that the patient alone would in all likelihood find difficult to examine, to navigate, or to ultimately manage, the therapy needs to be able to help the patient confront difficult issues in order for that person to ultimately struggle better. This means that everything will not be comfortable.

Thus, a new question is raised: Why is it the case that empathy can increase resistance rather than decreasing it? The answer to this question concerns how the patient feels about the empathy insofar as the patient, the recipient of the empathetic comment, will necessarily feel grateful. The feeling of gratitude necessarily means that the patient feels understood and safe and is thankful. However, this in turn also means that the patient's typical defenses are also safe. Because the patient feels better, and safer, and relieved by feeling understood, a claim can be made that under such conditions, one's typical defense posture and one's typical way of being is not at all confronted and the gratitude the patient feels is the mechanism by which all the defenses and typical self-defeating characteristics of the patient's personality get, so-to-speak, also to feel safe—from any challenge, especially a challenge from the self.

The conundrum therefore with respect to the therapist's successful expression of empathy (an empathy that is truly necessary for the establishment of rapport in order to sustain and then develop the ongoing therapy), also contains a negative effect of reinforcing the patient's sense that everything is alright the way it is—that is to say: "that I have an ally in my therapist." The problem is that such an alliance (what therapists refer to as "a therapeutic alliance"), is on the face of it seeming to be a good thing, and yet in the infrastructural underpinning to such an alliance lies a conflict, or contradiction, or even an oxymoron; that is, the gratitude felt in regard to the therapist's empathy is precisely what the enlightened therapist will understand in terms of how such gratitude empowers resistance by reinforcing one's typical defenses and character traits, in turn very likely enabling the patient to breathe a sigh of unconscious relief. The so-called sigh-of-unconscious-relief expresses the patient's feeling of being understood, again in addition to its very possible reassurance that everything "within" is okay the way it is.

Thus, empathy permits the patient an experience of ease. It enables the patient feel safe, and by virtue of the response it invokes, such empathy then has a predictable calming effect. Yet, empathy itself never goes to the heart of the issue. With empathy alone, virtually nothing of an in-depth therapeutic advance is made except to offer the patient safety and emotional security. And of course this in itself can be applauded. But as far as having far-reaching therapeutic effects in the sense of at least partially reconstructing one's typical patterns, or in approximating some deep understanding of the patient's problem or conflict, empathy alone can do none of that. As a matter of interest, what empathy consistently wants is for more empathy, and thus empathy can also be addictive—it can validate the individual in a way that offers security in the absence of tangible substance, safety without an in-depth therapeutic context, and emotional tranquility without permanence.

And yet ironically, empathy on the part of the therapist is no less than a vital factor in helping to create a rooting therapeutic process. Therefore, empathy in the therapeutic process is good and at the same time if over-calibrated can perhaps not be in the patient's best interest. Empathy can simultaneously both facilitate as well as delay progress, and so it can have a reasonably positive upside therapeutic effect with respect to emotional support of the patient, but also a negative downside effect with respect to impeding genuine therapeutic momentum.

Now, to God

The idea of empathy as it relates to one's prayerful worship of God or one's respectful deference and embrace of Godliness is also noticed by Wright (2009a) in his discussion of "reciprocal altruism" in which mutuality (especially of partners understanding one another) generates in these partners a proverbial sense of gratitude. This altruistic reciprocity is the model across species (especially higher-order species) of the intrinsic interaction of empathy and gratitude in which a mutually satisfying relationship is established between the interacting subjects. And along with gratitude, as Wright notices—obligation enters the picture.

God and the Worship of God

With respect to God and the worship of God, the question can be asked as to how this issue of empathy and gratitude as well as *obligation* relate to God and the worship of God? For at least a tentative answer it is important to see that a person's relation to God whether on a one to one basis, or whether based upon a community action of temple prayer is also a condition able to produce for each person a sense of security, safety, and emotional tranquility; a condition often or even typically referred to as a transcendent or spiritual experience. Such feeling is one of receiving (or creating) empathy along with a corresponding response of gratitude for the opportunity to be in such an affiliation—a Godly affiliation. In addition, an inevitable sense of *obligation* then completes the reinforcement of the emotional benefit gained as a result of participating in the entire prayerful context, also including the sense of affiliation with the goals of that context.

It may be interesting to see how this process unfolds; that is, how does one manage through prayer to acquire the feeling of having had an empathetic encounter that automatically produces feelings of security, safety, and emotional tranquility? One answer is that an accepted communion with God occurs as a result of such a prayerful process, and offers one (as a result of feeling understood), a palpable release from tension. It is simply the power of feeling understood—essentially it is a profound power (Kellerman, 2009b). However, this does not necessarily mean that a God exists and that the communion with such a God is actually occurring. But if the prayerful homage does not mean that that a God truly exists, then what can it mean?

The proposed answer is that in prayer, if a God is truly present, then it can be assumed that it was such a God that offered the empathy. However, if no God was present then it can be concluded that one is giving empathy to oneself. In such a case it is the self giving the same-self, self-understanding, love, and even forgiveness. In such a case, one is offering an empathetic response to oneself displaced as it were onto a presumed actual God. In psychoanalytic terms it would be a case of God as man's projection (or externalization), although not necessarily to be considered psychotic. The person can feel the empathy even though the actual psychological projective dynamic remains as an uncrystallized thought. In other words, the person can experience empathy but does not understand that presumably the empathy was invoked and then projected, extruded from that same-self person's unconscious. Freud (1927/1961) called such belief, illusory, and, of course, illusion is not necessarily psychotic.

Assuming for a moment that such communion with a God is a projection for which the prayerful person is solely responsible, it may be understood as a managed illusion because the person praying truly believes in the very respectful connection to this presumed projected God. The empathy sought by the supplicant (the prayerful person), and then assumed to be supplied by this God, enables the supplicant to feel great gratitude for this God's benign patience in listening. With the empathy received by the prayerful person, and the gratitude experienced toward the understanding God, the supplicant is then relieved of a certain tension, and can, and usually does feel more secure, safe, and frequently transcendentally tranquil. The subsequent obligation felt by such a prayerful person to abide by the assumptions of this wonderful subject/God relationship presumably also then acts to sustain this special relationship, and also further reinforces such a reverential communion.

And this is positive in the sense that people deserve to have a way of helping themselves with difficult emotions, problems, conflicts, and anxieties. However, whether such experiences, considered to be healing experiences, constitute a test or proof of God's existence becomes a rather conjectural or putative question. In any event, it cannot be decisively denied that God may in fact, actually be present, as psychological mechanisms such as "projection" could also simultaneously be at work. In fact, there could be two Gods present—one, the real one, and the second a projected synthetic or fabulized one. Of course, again, if God is in fact inexistent, then the entire prayer enterprise consists of a way of unconsciously offering empathy to the self, and then receiving it with gratitude thereby ending with a general sense of healing and relief of tension, and at least temporary dissolution of emotional agony. The other attendant issue here concerns whether or not such relief is actually one based upon permanent resolution of internal conflict or whether it is all based on "wish," and remains quite temporary.

The final factor that institutionalizes the entire process of prayer and gratitude to God (for God's patience, mercy, and forgiveness) is the reinforcement the subject has received throughout life from parents, teachers, family, and friends that validates the presence of God, and therefore supporting the belief in God's actual existence. In addition, a person's attendance in a House of Worship for the purpose of adoration and tribute to God, also provides a healing environment in which sermons and prayer reinforce and fortify personal belief—especially also because now there are many more people doing the same thing. An affiliation with such a House of Worship now constitutes a community that creates both cohesion of community value, or even an adherence or adhesion to the entire validating process of thinking and believing that it is the prayerful person's relation with an actual God that defines its salience. This means that such affiliation can lead to a super-allegiance. Because of this heightened allegiance, even responsibility to one's own safety can be forsworn in any sacrificial goal as long as it satisfies and even sanctifies the self by doing the presumed will of God. And in such cases religious extremism may lead to sacrificial suicidal so-called humanitarian acts, or heinous homicidal acts.

In contrast, and in this sense of adhering to an affiliational ethos, and with respect to a rather positive outcome, it is in the throes of worship and adherence to the belief in God that can erase a person's need for all sorts of delinquent, or self-defeating behaviors. For example, born again Christians have reported that the only way their alcoholism was cured was through their devotion to Jesus; personal anecdotal reports are of the spirit of Jesus entering the person and pervasively affecting all their behaviors. And apparently, this commitment and affiliation with Jesus is able to permanently triumph over any number of addictive or delinquent behaviors. This sort of transcendent experience is also reported by Muslim and Jewish believers as it relates to their respective objects of devotion. Thus, for believers, this kind of experience becomes decisive in their appreciation of an affiliation with God. For nonbelievers, such anecdotal references are often considered to be belief by suggestion; that is, a kind of self-induced, almost hypnotic belief—also infused with posthypnotic possibilities.

Appreciation for the reverential affiliation with God can apparently also be reinforced by what Nicholas Christakis and James Fowler call "social contagion" (Christakis & Fowler, 2009; Thompson, 2009). Social contagion is akin to an interconnective almost surreal appearing communication that is active among networks of people so that one person can be influenced by another who is not even in immediate proximity but who has communicated with a third person. It is that third person who then conveys the spirit of the message to another, and so forth. This is an example of the development of strong bonds through the reinforcement of various mediating conditions that fortify affiliation.

A negative outcome of strong adherence to an authority such as the authority of God, again, can also be cited as the reason for any number of cases in which people have carried out atrocities in the name of one God or another. In the study of affiliational psychology such cohesions, or even adhesions to a person's belief-ideology and corresponding adulation of a God have revealed how such loyalty—especially in its chauvinistic form—can lead to a kind of hypnosis (or even posthypnotic suggestion) that can be sustained even for a lifetime, and which subsequently can also compel people to engage in and enact atrocious behaviors.

In all cases (both with respect to the positive as well as the negative implications) it is a function of the person wish to be understood and receive empathy from the "Other" that generates a sense of well-being, safety, balance, and security—sometimes because of doing "good" and at other times because of doing "bad." Either way, acts are justified if they are perceived or felt to be God's will, and it becomes permissible to do whatever is asked. In fact, anything asked becomes quite possible to accomplish because all ethical imperatives will have been exchanged for the sake of emotional balance, sense of well-being, and need for safety and security. It is then that the individual actor behaves and believes that the hand of God is implicated. Under such conditions, one's moral compass is determined from the outside rather than from within; that for better or worse, one's independent thinking will have been compromised—in fact, most likely, usurped. Just as in psychotherapy where empathy wants more empathy without necessarily creating any change in the person's personality, much the same can occur either in communal or in individual worship of God. Typically, people feel better with respect to their worship, but still, many people proceed to commit their same troublesome behaviors, and then again return to pray in order to regain their sought-after relief from tension or pain.

Empathy and Change

Despite prayer's ability to help people defeat addictions and other self-defeating behaviors, prayer like empathy can become an end in itself occurring in the absence of effectively resolving conflict. And if by chance the conflict does weaken, it may weaken without any insight regarding what it was that animated the underlying problem or self-defeating dynamic that caused the problem in the first place.

Thus, prayer can be akin to what psychoanalysts call the repetition compulsion—the need to continue to seek empathy in any number of contexts outside the psychotherapeutic endeavor, including in prayer to God. Yet, in the absence of any working-through of conflict and therefore in the probable absence of possibly more permanent change and struggle for the better, and in the sense of sustained repetitive behavior, usually nothing changes—except, in most cases, the prayerful person in an homage to God within the context of prayer can also achieve a neutralization of anxiety and consequently an increase of peace of mind characterized by the warm embrace of gratitude.

The Power of Personality

Simply because one feels gratitude for the healing embrace of empathy does not make one receptive to change in personality patterns or emotional/psychological symptoms. The character pattern or pattern of behavior, or personality formation, or particular symptom does not at all listen to logic. One's character pattern and/or symptoms do not understand the language of logic. It is virtually impossible to "logicalize" psychological symptoms. Logic and even sensitive understanding can never change personality, or character patterns, or symptoms. Never! Of course, the exception is in the psyche's ability to create a system of wish fulfillment whereby for example, one's adherence to, and identification with God can begin to dictate what should or should not be done—and it might work. Wish fulfillment (and its vicissitudes) in the person's psyche is, to be sure, a very psychologically tenacious God itself.

That's the exception to the major impotence of empathy as a way of curing pathology; that is, empathy and its relation to the persons psychological system of wish fulfillment can impel the individual to do many things. One of those things is to relieve a person's pressure—which empathy does. Cure? No. However, as indicated, this kind of empathetic power is only a method to feel better. It does not touch anything that is repressed such as the underlying conflict(s) at the core of any of the person's problems. And without touching the conflict(s), whatever is unconscious, remains so.

Thus, without the surfacing, struggle, and possible resolution of this assumed underlying unconscious conflict, the associated amelioration of manifest anxiety through prayer reinforces the connection between such prayer and worship of God solely with a synaptic sense of feeling better. In a sense it is similar to classic Pavlovian conditioning.

Therefore, in this chapter, it is possible that an alternative paradigm of understanding prayer and its relation to a presumed God has been at least tentatively unfolded; that is, that a new dimension to prayer can be seen to contain a profound mental or psychological mechanism permitting one to unconsciously project one's need to be understood by such a presumed God in a nonpsychotic belief that such a communion is actually occurring. It could be considered self-fulfilling prophecy— you want a protector, so you make one, and in this way the needs of the person's psyche are never denied.

Assuming this is correct, then the need for empathy actually supplied through an ingenious human ability to express gratitude to oneself by psychologically engineering communication to a mind-projected God (with an absolutely certain result of understanding from this God) then, in essence is engineered by the self. In the sense of projection then, it is one's empathy given to oneself, and one's self as unconsciously transposed as God!

Again, it is possible that the entire prayerful homage to God imitates a condition of auto-suggestion as activated in the hypnotic process. In this sense, the prayerful person can be self-hypnotizing and even creating the condition for a posthypnotic suggestion fostering a concatenation, a string of emotional and psychological synapses leading inexorably to the need for more prayer.

None of it, however, has not the iota of evidence that an actual God does not exist; that is, that no matter the logic, the rationale, the finest psychological or even systematic discernment of ostensible evidence contrary to the existence of a God can claim absolute certainty in the absence of a God.

Reversing Evolution: A Way to Understand the Psychoevolutionary Discovery of God

Why is it that everywhere on earth, either idols are worshipped and endowed with special powers or that Gods are worshipped the same way? Does not that imply at least that everywhere on earth people come to the same truth about the existence of

God even though the cultures, societies, and conditions around the world presumably are so different? An answer to this question might become visible if we reverse evolution and see what happens. And what perhaps happens is this: if we reverse evolution we see that the cerebral cortex recedes and just about disappears and at the same time the function of the tail as a protective mechanism retains its usual point of reference. Correspondingly, we will see that as humans recede into remote ancestral beings, the relation to God that has offered so much safety, security, a sense of balance and well-being is now, to some degree at least, attained by the function of the tail. The tail becomes the gyroscopic mechanism that offers balance and security and safety. It becomes the eye behind the head rather than the eye in the sky that was believed to be God's location. The sense of safety and emotional tranquility is more assured by the swishing tail telling the animal that all is okay in the back, or by the tail that facilitates locomotion as in fish, frequently also enabling the animals to escape potential harm, or by its prehensile function of grasping and holding, or for social signaling. To apprehend even potential harm in a predatory unsafe world is what surely qualifies as excellent reality testing.

The sense of emotional security, physical safety, stability, and overall peace of mind is now certainly, at least partially, facilitated by the function of the animal's tail. And as we fast forward in our hominid evolution (our primate ancestry, including humans), to Homo sapiens (appearing about 300,000–500,000 years ago), perhaps we see that all of it (emotional security, physical safety, tranquility, and peace of mind) is, in the absence of a tail, and in the presence of a cerebral cortex (a thinking mind), now creating a mechanism that transfers the function of the tail into the "sky" in the form of a God that presumably watches over us. Therefore, it matters not where on earth a person is, or in what country or culture, it seems to be that the Darwinian need for survival creates Freudian defense mechanisms which in their operation acts as a way to then assure us that we are being watched over and cared for—that empathy for us (and fairness) exists and can be counted on "from above," rather than "from behind."

Assuming that God is a psychologically human cognitive construction (perhaps a classic example of colossal human grandiosity and single-minded wish fulfillment—constructing God means being God) considered for the good of the human being and based upon the subjective human experience that assesses the world as a dangerous place, a place where the person needs protection, then applying the same psychological principles, we derive the polar opposite of God—that is, the presence of the evil one who gestates all the danger—the Devil, or Satan.

It is unlikely that the concept of the Devil, or Satan, has been highly selected in evolution and then installed into our genes. By the same token, Wade's (2009b) proposition that religion (ultimately meaning of course, a turning to God) has been highly selected in evolution and then installed in our genes is similarly unlikely. Whether it is the dancing ceremony 9,000 years ago, or the time of corn-based agriculture about 1,500 B.C., or whether it is a reference to hunter-gatherer societies of 15,000 years ago, all of it although correlated archeologically with religious-like ceremony, is not at all a convincing argument to lock religion into the DNA of evolution. However, it is quite a convincing argument to propose that the ritualistic behaviors spawned by religious practice addressed *fears* and *needs for protection*, and it was such security operations regarding survival that were actually locked into evolution. Religion, as Wade proposes, can facilitate social cohesion, but with respect to Wade's position, at best, it would seem that religion is an epigenetic phenomenon (the effect of environment on genes), and not a genetic one (by biology alone).

As far as the position of religion and the worship of God as a hard-wired neurological brain mechanism with specific anatomical brain locations is concerned, various researchers have targeted this subject matter for discussion. Such discussion shall be reviewed in the following section.

Religion and Neuroanatomy

The function of specific brain regions account for the person's ability to experience an entire range of emotions including joy, and anger, and even the sensation of mystical awe. These brain structures have been used as arguments both for and against the hard-wiring of religious belief or as evidence regarding the possibility of an innate God "module." Neurologists V.S. Ramachandran, and S. Blakeslee, in a book titled *Phantoms in the Brain* (1998), point out that despite the fact that mystical and spiritual experiences occur universally, nevertheless this would not mean that there is a God region in the brain. In reviewing their work, Flaherty (2004) cites what these authors propose with the following analogy: "… while every culture knows how to cook… no one would argue that there is a specialized cooking region in the brain."

Flaherty further points out that religious experience can be invoked by a significant number of brain functions that together can comprise such a so-called God module; that is to say that when all of these regions of the brain are activated (temporal lobe structures such as the amygdala, hippocampus, and temporal cortex, as well as with parietal lobe activity) mystical experiences become completely possible. Of course, Stephen Jay Gould would see this as an orchestrated "spandrel" event—a synthetic event based upon the operation of primary organic brain structures. Flaherty also points out that such brain functions as they correlate to religious experiences are used by theists and atheists alike in their arguments either to prove or disprove the existence of God.

The Spandrel

The fact that religion or ritual occurred in societies throughout hominid evolution and ostensibly in every region of the world as well as in all societies does not at all logically mean, as Wade suggests, that such ritual or even God-belief was ever wired into out neural circuitry. Instead, an alternate explanation, as stated above, registers the point that it was the *need for protection* that was wired into such circuitry, and as a result, various forms of both tangible and intangible structures were created to implement such a protective need; ergo, in this case, the intangible—religion. And it is this intangible epigenetic structure, this turning to God that Stephen Jay Gould, the Harvard paleontologist, and Richard Lewontin, the population geneticist, first identified and defined as the "spandrel" (1979).

The spandrel is a term borrowed from architecture. It is a space created by a structure that has nothing whatever to do with the function of the structure. The example given by Gould and Lewontin, and a further discussion of the spandrel (Gould, 1997), was reviewed by Henig, in a New York Times Magazine article (2007). Gould and Lewontin compare it to the building of a staircase which entails extra space under the staircase that has no pre-intended purpose but could, for example be made into a closet. This is then an example to explain the complexity of the brain as one which creates many spandrels. In the case discussed here, this idea would be understood as the person's need for protection that presumably became genetically a hard-wired brain function enabling the person to be wary and vigilant in a predatory world, and then anthropomorphically the spandrel emerges as an animated embodiment endowed with the properties of the wish for a God. Thus, for Gould (1997), the spandrel is a side-effect phenomenon of a phenotypic characteristic rather than a function of an evolutionary adaptation. In other words it is something that is left over—an "exaptation"—serving a function other than for what it was originally adapted.

With all of this danger and evil in the world, and therefore the need for protection, it turns out that the human psyche is truly phenomenal, and apparently will not be denied its wish for protection, because again, it was the need/wish for protection that was probably phylogenetically wired-in, which then synthetically produced a spandrel, universally and mostly called, God. While this does not in itself refute the existence of God, the psychological phenomenon of "projection" (believing that what you wish by externalizing it and seeing it in the world—also called "projective identification"), is quite a challenge to believers of an actual God. In this sense, the argument that the projective psychological mechanism is at the basis of the belief in God, is quite compelling. If projection is what is actually at work in the worship of God then the person, because of this phenomenon of psychological projection, actually makes the self, God. In this way because of this phenomenon of projection, the basic spandrel consists of the self externalized in the form of God. The spandrel becomes an epigenetic phenomenon, meaning that certain genetic givens (need for protection), are modified by environmental variables to create a new form, a new protective form—God.

As a postscript, it can also be suggested that the need for protection and search for safety that worship in God invites (and satisfies), also correspondingly satisfies Freud's connection between his definition of the so-called pleasure principle (the need to achieve satisfaction), along with an emendation of his definition of the death instinct—that is, an alternate definition of achieving elimination of tension, usually thought ultimately to be achieved in death—the quintessential condition of zero tension.

Therefore, it can be assumed that according to the Freudian consideration of man's most powerful desire—to achieve a tension-free existence—worship and belief in a God perhaps would be man's greatest discovery/invention—the God spandrel that nullifies tension by eliminating the finality of death; with the presence of God, it is possible that to be with God means there is no death.

The entire process leading to a discovery of God and then the worship of God is relevant to this discussion because of the present psychoevolutionary perspective in which it is seemingly compelling to see that the tail in evolution, along with mediating variables, has possibly become with respect to function, transmogrified into the cerebral cortex, the thinking brain. In this sense such a process leading from tail to brain satisfies the connection between a "proximate" explanation (how it all works) with that of the "ultimate" explanation (why it all works), a formulation also discussed by Confer et al. (2010).

Further, as was suggested in the Introduction to this volume, the organism's search for "fairness" is also very likely wired into all of DNA. In a summarizing article by Natalie Angier (New York Times—Science Times, July, 2011), Angier states: "… evolutionary theorists say our basic egalitarian leanings remain." This means that fairness expectations are derivatives of earlier evolutionary life forms. Further, she points out that Darwinian evolutionary theorists feel there is "… the legacy of our long nomadic prehistory as tightly knit bands living by veldt-ready team-building rules: the belief in fairness and reciprocity…" As she puts it: "A sense of fairness is both cerebral and visceral, cortical and limbic."

In the homologous (functional) trajectory of "tail" to "mind" therefore, and in the sense of psychoevolutionary process, an homologous journey can be traced with respect to the adaptational functioning of organisms aiding essentially in the quest for survival. This homologous psychoevolutionary journey also samples the effect of culture on biology and is expressed by Confer et al. (2010) in their treatise on evolutionary psychology. It is an evolutionary psychological adaptational process involving the interaction of environments with psychological and brain circuits, such as the interactive roles of culture and genes. Alcock (2005) and Buss (2008) also provide validation for the principle that such adaptational processes are consistently seen in nonhuman animals as well as in humans and have direct survival implications. Again, with respect to such adaptations, an example is provided by Bracha (2004), who describes fear adaptation of freezing, fighting, and fleeing, as designed specifically to address such threats to survival.

Within the evolutionary context scientists have applied the heuristic use of homologous function from a variety of vantage points (Garcia, 2007; Muller, 2003; Stiedter & Northcutt, 1991; Wagner, 2007). In most cases homology is considered the study of parallel behavioral structures comparing one species to another, although in some cases it becomes the study of how a single trait develops into a more complex syndrome. Essentially this kind of homologous tracing displays a trait on an evolutionary continuum.

Apparently, in the quest to develop a more complete understanding of evolutionary process it is considered useful to utilize a homologous template in order to uncover evolutionary transformations of organisms. Homology becomes immediately relevant in this uncovering insofar as vicissitudes of homologous phenomena reveal forms or varieties rather of "functional homology." Garcia (2007) refers to such functional homology (and functional variation) in evolutionary cognitive science, and Lee et al. (1996), as well as Benjamin and Zschokke (2004), discuss internal cognitive mechanisms as subject to homologous "force." Love (2007), also in discussing homology of function considers the factors of "activity function" (what something does) with that of "use function" (what it is for).

All in all, in the transmutational journey of the security and safety operation (as one issue of the "activity function") of the animal's tail, to the appearance of this same security and safety function seen in the development of the brain/mind, in our present study here, rather than utilizing the definition of homology with respect to comparing different animals (as in parallel fashion), we are using it to trace evolution of function longitudinally. And of course, we are interested in security and safety needs that also include peace of mind, a search for fairness, and the general pursuit of survival.

Therefore, it could be suggested that one's search for God (a protective "parental" God is implied) is essentially sought because all living beings apprehend the world as a predatory and therefore dangerous one. In addition, with respect to the ever present abundance of random events, "fairness" as an expectation for each minute of living is obviously not at all very likely to be a wish that is consistently realized or gratified.

Enter, the "homological" God, the one who can make it fair.

Chapter 2
God and Ontological Anxiety

With the disappearance of the tail in the evolutionary process along with the gradual appearance of greater cortical development, the thinking brain became also an ontological one. That is, Homo sapiens of course retained its animal concern with survival but added to its existential awareness, that of an expectation and tension about death. And from this knowledge and expectation of the certainty of one's ultimate demise was born what became known as "ontological anxiety," defined as an existential concern regarding one's "being" (Kellerman, 2009a)—a concept philosophically discussed by Kierkegaard and also psychologically elaborated by Rollo May (1950, 1983).

Ontological anxiety is considered to be an anxiety or tension generated because the sense of one's ultimate survival possibility is instantly translated by the thinking brain into one's ultimate survival impossibility. It is simply the prospect of one's death—the ultimate disempowerment.

As awareness of the inexorable march toward the end of life gradually dawns more and more on the individual, (i.e., becoming gradually more conscious and therefore better articulated to the self), the issue of one's ultimate disempowerment creates a perpetual ongoing tension in the personality. The hard-core psychological principle that causes this perpetual tension can be stated as: *Disempowerment, or helplessness, always (without exception) generates (or gestates) anger.* Since anger is an inherently assertive response, then when one is disempowered, frequently the only way to become reempowered is by being angry. The reason for this adamant connection between disempowerment and anger is that the human psyche will not tolerate any sustained disempowerment.

It should be noted, however, that while sometimes anger will be expressed openly, and at other times mediated by socialization factors, anger will be suppressed (amorphously sensed, or only partially conscious, or repressed entirely—out of any conscious awareness). Nevertheless, and presumably, the psyche is considered satisfied even with a psychological reempowerment that is repressed and unconscious. It is the sense of reempowerment that contributes to the person's overall feeling of security and peace of mind. It is apparently what the psyche wants; that is, in reality, of course, wishes are not always realized, but in the psyche, no wish will be denied. The psyche insists only on security and empowerment, and to this end the psyche

H. Kellerman, *The Discovery of God: A Psychoevolutionary Perspective,*
SpringerBriefs in Psychology, DOI 10.1007/978-1-4614-4364-3_2, © The Author 2013

may be said to calibrate judgments of the brain in the skull as well as the brain in the gut. The psyche regulates an amalgam of survival mechanisms consisting of powers of cognition, instinct, emotion, and intuition, and therefore in a psychological sense, the psyche is entirely governed by wish-fulfillment needs. In the end, the psyche can only accept security and empowerment.

Death

Along with the need to achieve with peace of mind—homeostatic sense of security, a consequent expectation of safety, the fervent hope of emotional tranquility, and one's need for ultimate fairness as well as a gyroscopic sense of stability—people are faced with the implicit contradiction of such ongoing peace-of-mind hopes with the existential sense that there is really no option, that one is ultimately finite, and in the corporeal sense, quite finite.

In this regard, earlier forms of life can also be ontological; that is, primitive animals too are always alert to dangers and become protective when their survival is threatened. And this is true even of the amoeba. If the particle ingested by an amoeba is experienced as noxious the particle will be immediately ejected. It is this one-celled organism's tropistic survival mechanism that "knows" what to do. Other typical instinctive or unidimensional tropistic behaviors designed for survival purposes are those that are implicit as simple fight/flight behaviors. That is to say that to survive, and depending on the threat to such survival, some flee while others fight. In any case all forms of life may be thought of as existential in their experience of the here and now. However, ontologically speaking, it seems that only humans can precisely perceive, and be exquisitely aware of time elapsing toward the end, toward death. With most humans (but certainly not all), the inexorable end is usually not managed with equanimity, and yet, it seems that no one can escape it.

Or can we escape it?

And so, as with the need for empathy and feeling of gratitude for being understood, and cared for by an assumed greater power, the human thinking brain needed to construct, and in fact did construct something outside of the box—something new and extra that could possibly be a promise of sustaining the self indefinitely into the future. In this sense, being sustained indefinitely into the future essentially means there is no end. The sense of it is that this sort of unended future is more fair than the "end" would be.

Enter, God!

The Box

In order to solve a problem, sometimes mathematicians find it necessary to construct something hypothetical (some factor or variable) outside the parameter of the given problem in order to solve what needs to be solved within. In other words they construct a variable outside of the box in order to solve the problem within the box.

In our discussion, God becomes the factor outside of the box in order to solve the problem inside. And what is the problem that needs to be solved inside the box? The answer is that the problem inside the box, in our corporeal lives, is how to give ourselves some relief from the ontological anxiety (anxieties) of our lives that plague us all of our lives—the quintessential one of course pointing to what is usually considered the key existential issue—our individual demise—death.

With a convinced sense of, and belief in God, people can sometimes feel spared from intense existential ontological death anxiety, or at least feel assisted in managing such anxiety. Thus, the literal definition of death generally accepted by most people assumes rather a different meaning. It can be said that even with the idea of corporeal death as it is cognitively understood, this idea of corporeal death perhaps on the one hand can be somewhat denied or on the other hand, even better integrated in the personality (in one's thinking) primarily because comfort derives from the special experienced affiliation with God. In such a scenario, and for the believer, the end can subjectively become something quite other than the "end."

Thinking in evolutionary terms, not only is the process of what follows (the future) but phylogenetic beginnings are inevitably considered as well. And when endings are considered, ontogenetic process (from origin to development) is correspondingly also implied, so that "end" means "end." Therefore, in considering the past moving into the present and then into the future, a question can be posed: How did it happen that the evolutionary trajectory went from the anatomical tail to the thinking brain, enabling the human to manage death by at the very least, and for many God-fearing people, perhaps even neutralizing the phenomenological dread of such an idea?

The upshot is that the God variable can be useful when anxiety regarding "the end" is consciously experienced. Yet, at death's door, whether this variable actually helps becomes an issue of individual personality difference. However, before considering how the God variable influences individuals, let us look at how we got from the tail to the brain.

How Did We Get From the Anatomical Tail to the Thinking Brain?

Here we are holding in abeyance the consideration of an actually existing God, and rather considering phylogeny (evolutionary development) as well as ontogeny (human development), in trying to understand what happened in evolution to take us from more instinctive behaviors of animals (even those without any significant brain structure) to higher-order animals who live in groups with rituals and cooperative behaviors (and of course greater brain development), and finally to humans with a developed cortex, a thinking brain. This human brain is a highly cognitive one; that is, such a brain can accomplish all sorts of fantastic thinking feats. And despite the human capacity for primitive behaviors, nevertheless, simply because humans can create complex thinking edifices, we then attribute such positive qualities to people, and relegate pejoratives to the behavior of animals.

The idea of rituals in animal groups is quite fascinating because the group becomes a cohesive one in which all of its members "know" where they are and what their function is in the group, or their position in the group's hierarchy. For example in bee or ant colonies, the entire functioning of the colony is akin to what can be considered a virtual brain that determines the distribution of labor and power—who does what in the colony and why. Similarly in apes or even in dolphin troops, distribution of roles enables the group to become adaptive which of course increases its survival potential.

In a wide variety of groups, with respect to the behavior of group members, researchers have consistently seen phenomena best described as "reciprocal altruism." This reciprocal altruism is behavior among members that says: "You help me, and I help you, and because of this reciprocity, we survive better." It is what Wright (2009a), in his book, *God in Evolution*, analyzes with respect to the appearance or emergence of God in evolution generally, and in human affairs specifically, and it is what Natalie Angier (2011) refers to in her review of the "fairness" issue as a genetic tropism.

It is evident that such ritualistic organizations of individual animals in groups with each member "knowing" its role and function, rather gradually becomes a "precursor functionality" in the evolutionary march toward the development of the thinking human brain. In this sense, we may begin to see the outlines of a way to understand this evolution. In general terms correlations obtain with respect to the tail, group ritual, and the thinking brain—designed in evolution for survival—for peace of mind, safety, tranquility, and from signal behavior to language.

At this moment in the development of the human thinking brain, we arrive at the nexus, the point at which our thinking capacity is able to either apprehend or create the greatest power, the God power; the one that for many people can provide the needed empowerment and peace of mind. And all of it presumably develops because of the ubiquitous need to adapt to environmental circumstances—meaning that the "Power" has been found, enabling the conquest of all uncertainty (and certainly the conquest of the looming dangers of the here-and-now world). It is an issue that highlights the importance of understanding why or how belief evolved. The possible answer concerns the ability of the thinking brain to accommodate the individual's needs as well as for that brain to create other thinking avenues—even those for which the thinking target was not designed. This idea of the thinking brain as directed toward targets of consideration and ultimately to permutations of thinking can be applied in other ways as well.

Therefore, to the question of why belief evolved, essentially there are two basic theories: the first is termed the *byproduct theory* (Gould, 1991); the second concerned what has been called the *adaptational theory* (Sosis & Alcorta, 2003).

The Byproduct Theory—Here, belief evolved out of a spandrel. Henig offers the standard example of byproduct theorists—to wit: blood cells transport oxygen throughout the body but there is no advantage in the blood's red color. Redness becomes a byproduct of blood containing hemoglobin.

The Adaptational Theory—Here, belief is based upon primary benefits to the person due to the possible survival advantages of, for example, religious belief.

Correlation and Causation

Of course the origins of God-belief and worship are not rooted in with any profound theological tracts and exegeses. Tracing it back to hunter/gatherer societies and even further back in time, we know that the thinking brain developed gradually and that at its dawning, it was in high probability a syncretistic thinking brain—that is, the kind of thinking that confuses correlation with causation and in fact considers correlation and causation convenient to believe as synonyms, and even for all intents and purposes to be the same. This sort of primitive thinking is characteristic of a search for meaning as a way to feel better organized, and less fearful. Most of all such primitive thinking represents the kind of thinking that generates the wish for empowerment. And the wish for empowerment existed for the emerging hominid at an ontogenetic time, and within the context of a carnivorously dangerous environment—one that necessarily was experienced in a steady-state of conditioned or anticipated disempowerment.

This syncretistic brain, again, defined as one that equated correlation with causation can be understood by any simple scenario. For example, an earlier ancestor awakens to a bad dream while simultaneously seeing and hearing storm clouds in the distance so that one's rather fragile lodging is potentially threatened. At this point the temporal association of the bad dream with the onset of storms becomes: storms as a result of the bad dream, or the other way around. And from such primitive correlational thinking, a simple synapse to idol worship is born—the Rain God, the Sun God, and so forth. Such thinking in the form of worshipping a "Power" would then tend to diminish tensions associated with vulnerability and disempowerment because recourse to such a superior power, to God, might provide relief.

Furthermore, psychological experiments have shown that we all remember unfinished tasks more vividly than we remember finished ones, and perhaps similarly, we remember things that went bad more readily than we remember all the good things (Zeigarnik, 1927/1967). And of course, at the dawn of Homo sapiens emergence, the primitive mind would most likely also have been subject to this same phenomenon of remembering the bad events more readily, which in turn surely had adaptive advantages. It is akin to the child touching the hot stove and then always knowing not to go near it.

The problem was that at such an early time in hominid evolution very many things were unfinished and untoward experiences including feelings of dread were most likely the rule. With such a prevalent under-powered condition of life the need for idol worship or fantastical constructions that offered some measure of solace and peace of mind would have certainly, in our putative, erstwhile, and syncretistic ancestors, lobbied for space in the psyche—in one's fantasy life, and wish-system. It is also the wish for a more fair existence in the absence of disempowerment.

Cognitive Tools

With respect to the issue of the evolution of cognition in the developing Homo sapiens brain—progressing from correlational thinking to the compelling rational causative thinking of modern man, as well as in the organization of memory in modern man—we begin to see that what eventually surfaced was an assortment of cognitive tools, the most important of these being what became known as: *agent detection*, *casual reasoning*, and *the overall theory of mind*. These cognitive abilities enabled the individual to gain a greater ascendancy with respect to survival.

1. *Agent Detection*—In agent detection, the individual assumes a presence rather than deny it. Henig's example (2007) is of a caveman who sees something move. This caveman instinctively and reflexively "knows" that it is better to assume danger even if in the end it was just a leaf rustling in the wind. This point is also referred to by Hazelton and Nettle (2006) as "error management theory." Henig makes the point by indicating that if it was a hyena, then the caveman's survival potential would be reduced. Henig further asks: "What does this mean for belief in the supernatural? It means our brains are primed for it, ready to presume the presence of agents even when such presence confounds logic." Such agents belong, so-to-speak to the byproduct theory; that is, that belief in the supernatural is a byproduct of another hard-wired brain event. According to Barrett (2004), and in this same vein, religious agents are either people with superpowers that are synthetically endowed with the power to answer requests, or can have disembodied minds, or that can control us in the real world emanating from some other dimension.
2. *Causal Reasoning*—It is normal and natural for the human brain to be entirely reflexive and even helpless to a compelling cause and effect logic. Even in ancient times, effect was always searching for the cause and so as Henig says: "The ancient Greeks believed thunder was the sound of Zeus's thunderbolt."
3. *Theory of Mind*—Again, with respect to byproduct theory, a number of authors can be represented by Bloom (2004) who states: "… it is a short step to positing minds that do not have to be anchored to a body. And from there, it is another short step to positing an immaterial soul and a transcendent God."

With the development of these cognitive tools even remote ancestors (not too different from modern man) could make survival thinking more hopeful, and by high probabilistic implication they also surely needed moment to moment respite from all the ongoing anticipatory real dangers. Of course we eventually get to "modern man" (that quintessential oxymoron; that is, in view of vast human derived brutality, greed, terrorists, genocides, and so forth, then if it is "man" how can it be "modern"?). So, in examining historical sequence, we can see that idol worship has been discarded for the certainty of one God (or maybe two or three depending on the particular religion, orientation, or culture).

The anthropologist Scott Atran is also cited by Henig as positing the concept of "evolutionary misdirection." Atran (2002) states the spandrel eloquently: "Evolution always produces something that works for what it works for, and then there's no control for however else it's used."

Thus, it seems that we all have an innate ability for belief but apparently it is culture that forms the content of that belief—or as Barrett (2004) tells us—that the content can be: "whether there is one God or many; whether the soul goes to heaven: or, whether the soul occupies another animal after death." For some, such belief in a soul, heaven, and God concerns what these others consider a nonbelief; that is, atheists decline and actually repudiate what they define to be otherworldly notions—considered to be ideation based entirely on impulse and opinion with no basis in reality.

Thus, the "ending" dimension to life that becomes the province of termination—actually the curtailment of life—becomes for believers a transformation of the dance-of-death in favor of a more transcendent belief which as a peroration could be considered a Godsend. As implied, the corollary question relates to the nonbeliever insofar as a new question asks: How does this issue regarding the discontinuance to life—its cessation—impact atheists?

Ontological Anxiety and Atheists

Are atheists not concerned with death? Are they not worried about it? Do not they need the same reassurance concerning either an afterlife or some sort of even a proto-fantasy of a sustained existence? It might be said that generally atheists feel God is a "going concern" so long as people understand Earth to be the center of the universe. However, atheists generally see the Earth as another point in the cosmos, more or less simply related to all other points (Pinker, 2002). In addition, not all atheists are the same. And, further, there are some who, in place of a belief in God focus on a vast unity of the universe of which they are a part; in such cases anxiety about the ending of one's existence, at least ostensibly, does not seem to gain significant currency because of this sense of being a part of it all. Thus, for such people, it also may be an affiliative reassurance through a connection with Nature along with a respect for Nature that offers at least some peace of mind, though fatalism may also play a part.

Fatalism is for some a compelling philosophical stance that many atheists assume, and such fatalism naturally evokes a "come-what-may" attitude. It is the psychological equivalent of an ego-less condition. It is the understanding or belief that no one is special in the sense of being a God, or being favored by a God—and since the point of it all is that really no one from above is looking after you, then correspondingly the best one can do in life is to do the best you can—and that's it. After that, nature takes its course.

In the sense of this sort of fatalistic (although not necessarily negative) attitude those atheists who are clear about their fatalism are not likely to suffer from ontological anxiety, or perhaps do not suffer with significant ontological anxiety. Contrary to popular opinion, even in a fox hole, such people are most often, and in the vast majority of cases, not praying to God, and further, at death's door they are not likely to convert—they just move into the end. This relationship with the "inevitable" is not to say that atheists ignore the inevitable and do not at all fight the "dying of the light." Atheists like believers experience the same resistance against termination. Dylan Thomas locates this fight in all people. However, at the very descriptive end, the atheist narrative reflected in a communion with (or belief in) nature, and the narrative characterizing the belief of God-fearing people are distinctively different in philosophy although similar in effect; that is, they both offer the promise of easing death fears. Or, perhaps, it is more hypothetically accurate to say that atheists are allegedly not at all afraid of death; of course, "allegedly," becomes the operative term.

With respect to ontological anxiety, atheists are more focused on the Darwinian instinct for survival rather than on a so-called God gene so that the attitude of fatalism frequently becomes the balm ultimately enabling presumed better management of the "inevitable." Atheists usually do not equate "the end" with any excessive or extravagant definition of unfairness.

In the instinct for survival, in the organization of human psychology, and in the essence of the human psyche, are contained mechanisms that enable worshipful people to believe in a God, and such mechanisms also generate additional mechanisms that in turn switch-off such belief–possibility so that to atheists, denying what is tangible in place of a belief in the intangible—especially with respect to a belief in a supra-natural God—is not possible. According to atheists, worship of a God requires such mechanisms as denial of reality, suspension of disbelief, projection of one's needs onto a supra-human figure/object, and then identifying and affiliating oneself in a like-minded community of believers either worshipping alone or together, and in either case, in communion with God.

To theists, the atheist position is as inconceivable as is the atheist's questioning of how believers can believe. However, whatever the belief, the entire issue of believing in a God or as some believers would define it—as finding the existing God—can be analyzed within the context of epigenetic considerations—that is, the context of understanding behavior based upon genetic givens in relation to environmental conditions that can either invoke these givens, or fail to do so.

The Epigenetic Human

The question is: Are we solely genetic beings? The answer is gradually unfolding to reveal that the human being is rather an epigenetic creature. It is currently fairly certain that genetic endowment—directly from birth and even during gestation—is encountered, even challenged, by environmental demands. Fraley, Brumbaugh, and Marks (2005) make the point decisively: "The framework of evolutionary psychology dissolves dichotomies such as nature versus nurture, innate versus learned, and biological versus cultural." These authors also state that environmental pressures are effective at the phylogenetic level (in the process of biological evolution) and are even manifested ontogenetically (during the lifespan of an organism).

And this raises the question of our moral metal. That is to say, is being "good" inborn, or is being "bad" inborn, or is it more complex than that, meaning that there really is a cause and effect interaction between genetic givens and environmental influences that unfold during development? The answer to this age-old question is now, because of epigenetic understanding, finding its context. And this context includes environmental experiences that have the power to potentiate various genetic givens, and that these genetic givens await their environmental influences in order to be awakened—actually activated.

This epigenesis issue brings us to environment, environmental demands, and the influence they have in how we ultimately and derivatively feel and behave. Scott (1980) proclaimed that with respect to the creatures inhabiting it, the environment contains challenges or even "functional requirements" confronting all life "for the purpose of adaptation and survival." And Plutchik (1980, 2001) spells out in great detail the phenomenal idea, also supported by Scott, that there exist common and identifiable adaptive behaviors found at all phylogenetic levels—from amoeba to man!

These adaptive behaviors are considered prototype behaviors (basic categories from which derived behavior is modeled). In pointing out and enumerating these adaptive prototype behaviors, Plutchik elucidates the principle that the context of an organism seeking survival gains increased survival probability through an adaptive quest for a safe environment. This search for a safe environment is accomplished in a number of prototypic or typical ways. These typical ways involve behaviors that facilitate the organism's need to gain equilibrium through the ability and motive to explore even its microbiological world. For example, these prototype behaviors include the facility to engage in behavior of incorporation (taking in), ejection (excretion), avoiding predation (protection behavior as in fleeing), becoming immobile, or instituting stopping motion in the moment of unexpectedness or momentary disorientation, or attacking a barrier to a goal. In the face of pressures to adapt, the idea of "minding" (Wyers et al., 1980) pertains to the need that all organisms have to maintain equilibrium. The biological process in evolution relates to development governed by survival pressures and so even in the struggle to maintain equilibrium evolution manages to progress despite continuous intervening stages of evolutionary disequilibrium.

Although the assumption that an understanding of survival mechanisms in the form of prototype behaviors can be utilized to support the evolutionary biological theory of life (as for example, in the psychoevolutionary perspective regarding the discovery of God), nevertheless, Rizzuto (1996) insists that there is another perspective to consider (other than biological adaptation) when discussing belief in God—that is to say, if not an understanding of the discovery or even origin of God, then at least an understanding of the belief in God.

Rizzuto, points out that psychologically, the issue of God's presence serves the purpose of ensuring psychic equilibrium; that the agency of God helps the individual retain some modicum of love in the face of untoward experiences as for example, in the experience of abandonment. Also, this same agency of God helps to sustain one's dignity and hope when life does not cooperate with one's wishes (especially when conditions become intolerable). And Rizzuto further points out that for many people, God is also a companion.

It can be seen that in the psychoevolutionary sense, a thinking brain in concert with environmental demands and challenges leads to the person's constant growing edge of adaptation in a never ending wish for greater equilibrium. And this issue of adaptation also raises the implication of whether for the sake of adaptation, the natural selection of evolution "wants" us to be good in the same way that the presence of a God would, by implication, also "want" us to be good? In this respect it could be hypothesized that natural selection as the evolutionary engine has indeed selected us to be good—that is, to do good things unto others even without the remaining "… as we would have them do unto us." This ending may imply that natural selection favored being good (cooperative behavior creates better adaptation), nevertheless, this "goodness"-inheritance, awaiting its complementary environmental stimulus (in order to become activated), depends rather heavily on early historical experiences in life (influence of ontogenesis), as in a child's need for care, and corresponding loving behavior from parental figures.

The So-Called Evil Gene and the So-Called God Gene

This epigenetic complex of the child's need and the corresponding ontogenetic appropriate care given by parents is what is defined as early and continuing good nurturing and parenting. It is accepted as a premise that children who have been loved, understood, and well nurtured grow up with overall good intentions toward others and without so-called evil behavior—that is, this epigenetic process occurs correctly when appropriate environmental stimuli are present in order to enable developmental tasks to be achieved at age appropriate phases of development. It is when bad behavior is everywhere in evidence that a belief in an evil gene is contemplated, and further, such bad behavior invites the belief that an actual evil gene exists. The truth is, or the hypothetical truth is, that there is no such thing as an evil gene! The evil of behavior and underlying intention can be understood to be a result of an obvious epigenetic failure in development; that is, it can be hypothesized that with respect to cruel or delinquent behavior, the calibration of genetic givens with appropriate environmental stimuli (as well as time of response) did not coordinate well, or at all, or that because of some genetic anomaly the person is born with an absence of whatever is biologically required to restrain impulse. In such a case it is the untoward amount of impulse that increases the probability of inappropriate behavior, not an evil gene!

When a child feels unloved, emotionally abandoned, or even abused, the "goodness" intended by natural selection because of adaptational considerations of such "goodness" cannot be normally crystallized, and therefore in the present evolutionary state of human society (a society that again, makes necessary the pronouncement of the oxymoron, "modern-man"), the adage about treating others in a positive and supportive manner actually does need the reminder, "as you would want to be treated."

Because of the frequent disconnect between proper parental guidance and subsequent normal child development, in their day to day lives, people need assistance in asserting and practicing a moral, ethical stance, thus making the particular suffix "as you would want to be treated," a necessary and even highly practical precept. More specifically, apparently, because of the rather primitive existing understanding by vast numbers of parents regarding developmental issues of childhood, and the colossal underperformance of what should be good parenting, it becomes evident that even though we know that "goodness," kindness, altruism, and cooperation are all ultimately better able to facilitate adaptation, we also know that reminders for such "goodness" are most definitely needed, and so the suffix "… as we would have them do unto us" becomes highly important in the struggle toward more decent behavior, and subsequently, in the unfolding of a more decent society.

It seems, at least hypothetically, that human nature "wants" to be good. And that such nature is only bad when the individual, during development, experiences varieties of depredations such as isolation, or serious emotional deprivation, all the way to abandonment (physical as well as emotional), and abuse. In such cases, one's wishes and impulses are not at all cohered and/or controlled so that adaptive socialization becomes contaminated and essentially arrested. In the absence of proper parenting, or even in the absence of any parenting, the chief emotions associated with the impulses of personality will be those that sample the aggression dimension: at the less intense range, annoyance and irritability; at the middle range, anger and indignation; and, at the greater intensity—rage, fury, and wrath. The point is that when one's wishes are almost never addressed, when one is almost never understood, and when one is unloved and uncared for, then the hard-core psychological result will be the continual expression of aggression, anger, and rage, and any variation of what becomes unrelenting claims of injustice. And none of it has anything at all to do with an evil gene. This idea of an evil gene or a jealousy gene is also contraindicated by Confer et al. (2010) in a discussion of adaptation and evolutionary psychology.

What this means is that even though "goodness" is apparently highly selected in evolution, nevertheless, the person still has the potential to be negatively inclined ("bad") because of the absence of necessary environmental stimuli that trigger genetic markers to produce the epigenetic natural result of "goodness." And such contamination constitutes the recipe for what is usually discussed as "bad behavior" leading some to believe that an evil gene in fact, exists. The question then can be asked as to whether a God gene exists. And the answer is that such a so-called God gene probably goes by another name. And this other conception representing a God gene (in the genetic makeup of the individual) can be referred to as "survival instinct."

Because of our survival instinct born of existential fragility (in addition to the apprehension and fear that people have of the "end"), a common solution is to seek emotional balance, and safety, and because such things are never easy to secure and rather always subject to unforeseen circumstance, what is generated instead is what we are referring to as ontological anxiety. It is an existential tension about keeping the physical corporeal self whole, and the emotional self, safe—especially in the face of an unsafe and clearly unfair world.

   Such prospects leading to physical and emotional safety as well as overall peace of mind require an empowerment—one that can be relied upon. Such empowerment is, of course, the belief in an all powerful and in most cases, forgiving God. For so-called lower forms of life this existential sense of insecurity and the corresponding instinct to protect against the presence of predatory intensions relentlessly lurking in the environment prompts the animal to utilize its tail (among other anatomical and sensory organs) as the eye in the back of the head—as anticipatory protection. In higher primates, identifiable ritual may share the analogous homologous function of the tail because ritual will assure a cohesive and relatively peaceful society with good adaptational possibilities, as well as reasonably ensuring a protective perimeter, a more secure circumference of the group.

   With Homo sapiens, it seems that belief in God serves the same purpose. Interestingly, nonhuman animals will usually concern themselves with the vicissitudes of survival but minus nefarious wishes that human beings exhibit, unless a so-called nefarious wish increases advantage. But of course, other than predatory hunger-seeking targets or instinctive species to species antagonism, nonhuman animals do not become serial killers. In humans, such nefarious wishes and associated behavior that people demonstrate are acted out in what is defined as evil behavior, and this sort of so-called evil behavior is not in the least, and strictly speaking, not even at all related to objective normal survival needs. Rather, such skewed and abnormal behavior is all designed to also satisfy survival wishes, but of a perverse nature—whether or not in the presence of any type of God. As stated, this kind of delinquent behavior can be attributed to failed epigenetic connections; that is, environmental stimuli did not calibrate well with genetic givens.

   And here is the logical paradox and psychological conundrum: People can and do participate in nefarious acts that have nothing whatsoever to do with survival, while at the same time such people even can be quite devout and very strongly believe in God. A prime example arguably might be the most destructive spy/mole in the US history—the FBI high ranking agent, Robert Hanson. The question is: How can that happen?

Which Are the More Powerful, Psychological Mechanisms, or God?

The question of which is the more powerful, psychological mechanisms, or God (meaning any definition of an objectified God or one defined by man), ultimately has relevance also to alternate propositions or questions. These are

1. Did God create man in His own image?
2. Did man create God in *his* own image?

Let us begin to examine these alternate propositions by looking at the case of Robert Hanson who is considered by many to be the person who perpetrated the worst intelligence disaster in American history.

Hanson was an FBI agent, who for over a 20 year period, used his insider knowledge to transfer highly classified documents to the Soviets—doing it for financial gain and because he was angry about not being sufficiently recognized for his ostensible talent and presumed value within the FBI. His clandestine transfer of information caused a series of infamous murders of American "assets"—counterespionage agents stationed abroad. He was finally discovered and being spared the death penalty, was given a life sentence. The staggering fact was that Hanson was a devout parishioner of the highly conservative Catholic organization, Opus Dei, and no matter the extreme points of view of Opus Dei, nevertheless this organization is not known for any viewpoint condoning serial killing.

The question arises of how a God-fearing conservative Catholic could at the same time be a person who would bring harm to others. Psychologically this can be explained by the idea of "compartmentalization" wherein two opposing viewpoints can be contained in one person, that is, the clean-cut All-American, God-fearing Christian, Robert Hanson, and simultaneously Robert Hanson, FBI spy/mole responsible for the deaths of many people, who compromised his country, acquiring a reasonable degree of wealth because of it—and then went to church!

And thereby also exists the explanation for that other creature of horror, the BTK serial killer, Dennis Rader, who was for 30 years a member of the Christ Lutheran Church, serving as President of its Congregational Council. In one compartment of his psyche, Rader took pleasure in strangling women to death, while in the other compartment, and simultaneously, he participated in sustained devotional supplication at his church.

Several questions can be posed in light of these paradoxical details. For example, is the psychological defense mechanism of compartmentalization (implying also the presence of other accompanying constituent personality defense mechanisms such as repression, regression, splitting, and denial) more powerful than the influence of God? Plain and simple, which is more powerful? It seems that Hanson and Rader, as proclaimed God-fearing men were completely compelled by a perverse psyche and not by the influence of God. In their cases therefore, psychological mechanisms were more powerful than God's influence. In other words, due to incessant and tyrannical inner impulses, these men were not amenable to God's influence.

Second, can we assume that the early developmental history of each of these men was characterized by experiences of deprivation, emotional privation, and absence of "goodness," and/or safety? And, third, is it possible to say (in a counterintuitive sense) that doing the wrong thing was exciting and valued because it actually provided comfort by giving each of these individuals a sense of achievement as well as the triumph of a comeuppance toward specific others who did them wrong—symbolic though it all may have been?

Therefore which is a more powerful, psychological mechanisms, or God? If God made man in His image, then God should be more powerful. However, it can be seen that in these cases psychological mechanisms trumped the so-called power of God so that it seems rather that it was man who made God in *his* image—a result of the psychological mechanism of projection; that is, projecting onto another qualities from your own wishes that for all intents and purposes usurp the personality.

An additional question becomes: What was it that caused these men (Hanson, Rader) to need the comfort, or excitement, or feeling of triumph leading to their nefarious activities, while simultaneously seriously worshipping God? It could just be that this kind of despicable behavior required, for these individuals, a furtive art. For such individuals, a church involvement coexisting with pernicious behavior becomes a form of expurgation of wrongdoing. Along with a need for the expurgation or sanitization of evil deeds exists the final touch of art related to such heinous behavior—meaning the reason(s) for the evil deeds which perhaps exist as unconscious motives. These treacherous and murderous acts become a final solipsistic curtain call—in church no less—for a job not only well done but artfully crafted with the savage and unspeakable scene left so that its creator remains anonymous. Therefore the art of it all is not only simply artful but also smarter—much smarter than us all. The evil, foul, malevolent, and depraved behavior which such individuals generated is almost certainly considered by such miscreants as choreography.

Thus, here we see a countervailing infusion of predilection; that is, we see a paradox of the coexistence of evildoing and worship of God. Such countervailing inclinations invoke inquiry into what such people who are pious supplicants (engaged also with serial killing) understand to be the type of God they count on. Of course this becomes a larger inquiry into the entire taxonomy of the classification of God with respect to God's nature; that is, what is God like? Is God powerful or not? Does God want to be powerful? Is God as powerful as a person? Does God care about people or not? Is God only interested in good communion? Is it an intervening God, a non-intervening or impartial God, or is it an irrelevant God? And further, is it an inexistent God? And even further, did God create man in His image, or did man create God in *his* image? And then actually finally, is God beyond knowing?

In the following chapter, these issues of identifying implied prevailing and characteristic approaches to understanding God, and the possible nature of God, shall be further examined.

Chapter 3
The Personality of a God That Can Be Possibly Known

Gods

Sigmund Freud in his exegesis on religion (1907/1959, 1927/1961) focused mostly on what he considered to be the illusion responsible for religious belief, especially the belief in a super/supra-natural God. He theorized that belief in God was a first step in a general attempt to manage the unfeeling nature of nature, or as Paul Bloom (2005) puts it, "the terrors of nature." In a specific way, Freud pointed to man's fate of ultimate death as a particularly onerous contemplation, and posited that religious faith was conceived by man to approach this harsh truth. In essence he saw it as a compensatory balm for this particular anticipatory (and unfair) existential suffering of human-kind.

Along with such a view of creating a God image, Bloom also cites Susanne K. Langer (1967) who stated that man has a very poor ability to manage chaos and that supernatural beliefs organize such chaos so that out of disorder and entropic disruption emerges greater coherence and even meaning. Such meaning generates emotional safety and a measure of psychological reassurance. Thus, for believers, with the prospect of achieving meaning and purpose and especially with the possibility of attaining eternal life, there was, it seems nowhere to have gone but toward God—the God that will make everything fair.

In this respect, the Freudian position on the genesis of religion and belief in God concerns the logical proposition made by secularists. Both analyses leads to the conclusion that such beliefs are not solely learned attitudes but rather a natural evolutionary product of the human mental structure—the thinking equipment enabling man to conceive of God, and that therefore this man/God relationship is derived from and emerges logically out of human sensibility.

The main point is that man encounters God because the search for security and safety invites the need for such an encounter. And, of course, the God that is encountered needs to be an understanding one who empathizes with man's search for understanding, appreciation, protection, and security. This need for security

H. Kellerman, *The Discovery of God: A Psychoevolutionary Perspective*, SpringerBriefs in Psychology, DOI 10.1007/978-1-4614-4364-3_3, © The Author 2013

presumes the need for tension-relief from the concept of death. The idea of a "soul" serves the purpose of negating death as a final end. Thus, for believers, there is no end. Therefore, the construction or discovery of God (whether artificially construed by man, or encountered by man and perceived as real) becomes a rather profound idea that can be used to achieve a sense of personal physical safety, emotional balance and security, and finally, peace of mind.

A most positive perspective of a belief in the importance of God is made by Michael Benedikt (2007) and referred to in Silver (2008). Benedickt says: "God is the good we do. Where we do not do good there is no God. Where and when we do good, there is God. God and our good deeds are the same." Of course a case can be made that such a conception leaves it up to the individual as to whether God exists and that therefore the implication can be that perhaps God is made in the image of man and not the other way around.

At almost the other end of the discussion as to whether a God exists is the comment made by George Bernard Shaw. Silverman (2006/2007) quotes Shaw: "The fact that a believer is happier than a skeptic is no more to the point than the fact that a drunken man is happier than a sober one." The really and truly other end of the God debate is one represented by the physics Nobel Prize winner, Steven Weinberg, who postulated the theory of "The Big Bang," and who said: "There will always be good people doing good things and evil people doing evil things. But for good people to do evil things, that takes religion" (Silverman, 2006/2007). An entire treatise on this issue of how religion has evolved is organized by Nicholas Wade (2009a) in his book, *The Faith Instinct: How Religion Evolved and Why It Endures*.

In face of this range of thinking regarding religion and God, Rizzuto (1979) states that no matter what the position regarding belief in God (implying also in the practice of religion), all people have some God representations whether or not a belief in God exists—a discussion of which is presented also in Aron (2004).

With this introduction to some of the thinking about the nature of God (and virtually, on the nature of man), the following will examine various conceptions of God and the possible motives and attitudes inherent in these conceptions; that is, what kind of God is it that we define, and what are the personality and ethical implications generated from such definitions? It is a treatise reflecting discrete differences in the kinds of Gods discussed. It is also an examination of the possible personalities of God(s), of course assuming that God is possibly "knowable." This is not Paul Tillich's God (1963), who is essentially beyond knowing. Because Tillich's God is also "being itself," such a God is therefore of the rather esoteric definition (nonpejorative), beyond essence and existence. Tillich's God is therefore not only beyond existence but is greater than existence in contrast to the examination of the "knowable" God. This idea is integral to a discussion of the discovery of God from a psychoevolutionary perspective, examining how psychosocial survival needs relentlessly impinge on evolutionary considerations.

The following is an examination of the possible nature(s) of God and how such nature can possibly unpack the meanings inherent and implied in the "psycho" of the "psychoevolutionary" perspective to the discovery of God.

As mentioned, in Freudian thinking the possibility of the reality of God is a natural byproduct of the dynamic of the human mental apparatus. This mental apparatus or mental structure comprises the physical brain (its neurobiological organization), the corresponding abstraction of mind (the cognitive or thinking ability of the brain), and the psychological construction of the psyche (the emotional/psychological infrastructure of the personality). With the quest for greater certainty considering all the uncertainties of life, the thinking brain and the subsequent awareness of threat to survival is animated so that an individual is predisposed to search for something to count on, just as the animal, in part, as a guarantee of protection from unseen threat, counts on its tail as the eye behind the head. And so, it is quite possible that with the tail no longer protecting the person as the eye behind the head, then with the thinking brain, people look to heaven as the protective eye from above—the eye that that is endowed with the power to right the wrongs.

J. Louis Martyn (1979), in an imaginative and even profound conceptualization states that one needs to understand God with stereoptic vision; that is, that God created heaven and earth and that even though dramas are occurring in both places, nevertheless, these are really a single drama because the story from above as well as the one from below are taking place on the earthly stage and are therefore, in an epistemological sense, the here and now along with what is to come. Therefore in the sense of God-created dramas it could be imagined that such a God is possibly or even probably in charge of events.

Enter the Intervening God.

The Intervening God

The Intervening God is the God who is prayed to often for an answer to a request, a wish, or a hope. And it is such an Intervening God that most prayerful people actually believe in, and for all intents and purposes, even insist on. And in the major religions (Christian, Islamic, Jewish), most believers understand that such a God is responsive to prayers for things that people want, or want to happen, or do not want, or do not want to happen. And in these religions, God, in turn, is not shy about asking of believers (in what could be considered an unstructured conversation) that they listen and obey. For example, God said to Abraham: "Obey me and I will protect you." Or, God said: "I am the Lord your God who took you out of the land of Egypt." In addition, Jesus instructed his disciples to do certain things and on how to reach "the Father." The noun "Islamic" translates as "submission," implying deference to God's wishes. All in all, these are reference to a God that intervenes.

And despite the intention of the Intervening God as a good God, and as a forgiving and friendly God, such a God nevertheless demonstrates a personality that can be severe and even punitive and that can generate a psychology in followers not just of rewards for obedience and punishments for disobedience but for the birth of indignant righteousness and its rather nasty accomplice, rage. Thus, despite the call for kindness and forgiveness, the Intervening God seems actually also to be capable of being a punitive one. For example, God instructs Abraham to sacrifice his only son. Human sacrifice (especially a cruel instruction as in the murder of one's own son [infanticide] to validate someone's allegiance) is simply not only cruel but also singularly primitive, if not strictly speaking, crazy. And although many have devised acrobatic explanations to justify and legitimize Abraham's dictum (also implicating God as accomplice), again, strictly speaking, no matter how you turn it, crazy is still crazy. Of course this Abraham/God story is also considered to be instructional with respect to a test of faith. Yet, it has the ring of god-awful, unkind, nasty, and even ugly judgment as a choice example by God to test one's faith.

Similarly, with respect to the symbolism of the Eucharist, of Christians drinking the blood of Jesus and eating his flesh, this symbolism may reflect a residual primitive ideational construction which millennia ago may have served a utilitarian religious and believed-for survivalist purpose. Yet, in contemporary life this sort of practice should already be vestigial. It represents a transubstantiation of eating human flesh and drinking blood. It really seems to be, in vivo, a symbolic cannibalistic practice, to wit: "This is my body broken for you—eat of it [the bread], and this is my blood shed for the remission of sins—drink ye of it [the wine]."

Finally, with respect to traditions particular to certain Islamic practices, the historic tradition of chopping off someone's hand as punishment for stealing, or stoning someone to death is also, to say the least, quite severe, and such practice is given a religious rationale that becomes an accepted legalism for those practicing it.

As implied in the foregoing, all of these examples raise important issues regarding the Intervening God. For religious Jews, one must ask about both of God's commands to Abraham ("Obey me and I will protect you" and, the command to sacrifice Isaac) as to whether one thinks that it is really Godly for God to negotiate, to make deals (obey me and I will protect you), and to challenge one's loyalty by forcing impossible requirements (sacrifice your son). Of course, no decent God asks for sacrifice of life, and the tortured logic about the impossibility of knowing the mind of God that some propose in order to justify such un-Godly sacrificial requests is with respect to intelligence quotient and rational thinking, nothing less than ridiculous. Even God's use of the verb "obey" is seemingly un-Godly because in place of such an un-Godly edict, would not a true good God rather proclaim the verb "think," or something Godly like that, instead of "obey?" Of course, from a psychoevolutionary perspective, compliance becomes an element of bonding which in turn also is directly part of the process of group formation—all of it facilitating greater survival potential so that in an almost skewed sense the concept of "obeying" could have psychoevolutionary resonance.

And how often does someone pray for something rather inconsequential and then thank God for granting it? One anecdote was that a woman prayed her husband would purchase the red automobile she wanted rather than the green one he wanted. When he surprised her with the red one, she said she knew God heard her. The answer could be that no decent God answers such prayers but lends a deaf ear to a mother praying for her child's recovery from a serious illness—to no avail.

In another anecdote, a woman (African-American) told a story about her childhood in the south of the USA where one night a gang of the Ku Klux Klan attacked her home with torches and rifle-fire. She hid under the bed and prayed to God to save her and her family. Luckily, these night-riders left without physically harming anyone. The woman said that such an experience was proof-positive of God's presence. However, the Klansmen rode off down the road and did the same thing to a house in which lived another black family who had a little girl praying the same way for rescue, but in that case, the Klansmen, burned it all down killing everyone in it including the little girl. The question to this woman, who when as a child, survived this ordeal, is: "Do you think God loved you who lived, but not that other little girl who died? Or, do you really believe that God loved you more than he loved that other little girl?"

Well, what do we make of this?

(a) God does indeed answer prayers and indeed intervenes in everyday life; or
(b) That if God answers a prayer for the color of a car but does not respond to the prayer of a mother pleading for the health of her child, then –
(c) Either there is no God because of this obvious tilt to the ridiculous; or,
(d) That God does in fact exist but we can't tell what his motives are; or,
(e) That God exists but He can't tell what His motives are; or,
(f) There is no God.

I believe it is interesting to consider choice [e]—that if an intervening God exists such a God certainly needs guidance because this intervening God cannot really distinguish right from wrong, or, good from bad. Thus, it could be (if choice [e] is to be examined) that God may be young/immature or even perhaps disturbed, and so needs guidance from people.

In other words, the intervening God needs to know that doing good things should be the instruction in the here-and-now life. Kindness, compassion, generosity, and education should be valued, while praise to a King or a God should not be the central concern of people. With this decent sort of sensibility, ultimately people will know that it is actually good not solely because God loves it, but rather, God loves it because it is good. Therefore, if it is an intervening God to whom devout people pray, it should be one who intervenes to prevent torture, murder, pestilence, hunger, and poverty. The final Godly intervention might be to eliminate the structure in the world that makes for the terrible drama played out by predator and prey. However in this unsafe world, Wright (2009a) suggests that more primitive cultures invoke the Intervening God(s) to explain why bad things happen (as well as why positive things happen). The prevailing belief is that it is all in God's plan and that the intervening God is one that connects to the people's need for greater control over their environment. These believers feel that such a God can obtain (or create) greater fairness in life.

Finally, Woodward (1992) alludes to the presence of the Intervening God when he points out: "… petitioning God for favors is one of the oldest and most human forms of prayer." And further, in the Gospel of John, Jesus himself promised his disciples that: "… whatsoever you ask the Father in my name will be given to you" ("ask" is the operative term and "given" completes the exchange).

With such conceptions, the Intervening God rules. The overwhelming conclusion, however, is that given the vicissitudes of human experience in the struggle for existence and survival, such an Intervening God needs to be very careful in establishing an ethical, just, and overall good Godly stance—which it seems has not been the case because universally people continue to *unsuccessfully* seek fairness in order to ensure security.

Just as in evolutionary development where transmutational events occur, so too in psychological development such processes also occur. This means that in a psychoevolutionary perspective of God, metamorphoses take place both with respect to the psychology of the person as well as in the biological sphere of development, and in both respects, reflect on man's encounter with the idea of God, or with God.

The Impartial God (Non-intervening)

The Impartial God, or the one who is non-intervening, is the God that looks and sees—watches, as it were—but remains neutral. Generally speaking, religionists consider such a God to be one who settles scores either in a way other than with immediate punishments, or somehow and in someway in the afterlife. Thus, prayerful people consider that asking favors of such a God is not relevant and therefore not appropriate. Rather, such prayerful people feel in communion with this impartial God and report their concerns, hopes, and wishes to such a God without expectation of response, feeling that the very act of prayer is an act of homage and not one of request.

However, the issue now becomes one in which a question arises as to the nature (or personality) of such an impartial God. Similar to the Intervening God, and with respect to the same vast cruelties and injustices in the world, can such a God, in the face of all this entropic and endemic lawlessness, remain at all neutral (impartial and non-intervening)?

This Impartial, Non-Intervening God, ostensibly made the world, but because for man, such a God instituted free will, the implicit Godly promise was not to intervene. With such a God, the Godly law forbade God's intervention no matter the persuasive arguments for such Godly intervention despite continual occurrences of worldly horrors. And of course there are people who pray to such a God in order to be spiritually connected and to register with Him the sense of affiliation and reverence, and not usually generally asking for favors or specifically for the granting of wishes—even though the seemingly natural urge to ask, still in all, and for sure, insinuates itself in the underlying inclination of the person's prayer.

This Impartial God is also noted by Eagleton (2009) and Wood (2009). Eagleton strikes a universal chord for the presence of God by stating that God is anything and everything "… the condition of possibility of any entity whatsoever. God is impersonal, not a celestial engineer … but an artist and aesthete who made the world with no functional end in view." Wood states that "God is for Maimonides not parental but self-sufficient…" For Eagleton, "God is not neurotically possessive of us." In other words, God is not the intervener. Rather, God is the power that permits the individual to be true to the self; "He is pure liberty."

Since, according to such conceptions, God is not neurotically possessive of us, then whatever happens to us may not necessarily affect God. And this kind of idea truly raises the issue of God's rather neutral position. This issue of the neutral impartial God can be illustrated by the example of a war-time photographer in an Asian country who was permitted to photograph the interrogation of an 18-year-old prisoner. The prisoner was about to be shot and the commander looked to the photographer (who had agreed not to interfere no matter what happened during the interrogation). However, the commander looked to the photographer and gestured: thumbs up (spare him), or thumbs down (kill him). Although instantly conflicted, the photographer decided not to abrogate the agreement and therefore chose not to respond. The request in gesture was again made, and again ignored. The soldier was then, on the spot, summarily executed.

The photographer was the Impartial God who as Eagleton says "… is not neurotically possessive of us," and is what Wood considers the God of pure liberty. This means we are all on our own because God will not intervene nor in any way compromise such individual liberty (freedom). Thus, "fairness" in the sense of God's neutrality cannot be expected.

In this particular example, the Impartial God, the one who believes in liberty, at least by assumption, becomes an ally correlating to the prisoner's inalienable right—to be shot.

The Irrelevant God

Robert Wright (2009b) just about defines the Irrelevant God even without referring to such a God as specifically irrelevant. Wright states: "In modern theology it is assumed that God did his work remotely – that his role in the creative process ended when he unleashed the algorithm of natural selection…" The idea of the God who did things remotely suggests a God that was unrelated or even uninterested in Earthly concerns. Rather, such a God was focused on the task-specific process of creating worlds remotely. And this raises the question of whether such a God has any value or relevance to people who worship? Does anyone really worship a God who is only remotely, and in addition only originally, and only technically connected to Earthly concerns?

The answer probably is that there are those worshipful people who are so focused on their own concerns and anxieties that their prayers and wishes are generally expressed with no idea of where and to whom they are being directed. Thus the irrelevancy of this Irrelevant God is really not of relevance to such a prayerful person because the only objective of such a person is to release the wish or prayer itself.

The Irrelevant God is for all practical purposes really the Anxiety God who is diminished by his own remoteness. The worshipper redefines this God so that a metamorphosis takes place making His function simply to reduce the worshipper's tension because the worshipper simply feels better by communing with whatever this God has become, while not expecting anything in return. Such a worshipper may even be content were the God to be revealed either as an Impartial God or per chance, even as an Irrelevant one. Yet the specific definition of the Irrelevant God is one not solely defined as a God that creates universes remotely, but really as a God that creates universes and then goes on to do other things and is not at all concerned with people. The crucial issue here seems to be that the absence of such a God translates as the absence of fairness or concern with people.

As we advance from an examination of these various implicit conceptions of God we are actually moving from the God who is more hands on to the one who is entirely hands off, and then to the one who is essentially inexistent.

The Inexistent God

Simply put, the Inexistent God is the God that does not exist. And of course there exist a proliferation of books explicating this position (Dawkins, 2006; Dennett, 2007; Harris, 2004, 2006; Hitchens, 2007; Kennedy, 2000; Pinker, 2002; Stenger, 2007). To elaborate this point, Graham (1975) in his book, *Deceptions and Myths of the Bible*, states: "It is time that scriptural tyranny was broken that we may devote our time to man instead of God, to civilizing ourselves instead of saving our souls that were never lost in the first place." However, at the other side of the table sit, Armstrong (2009), Atran (2002), Eagleton (2009), Haught (2008), Johnston (2009), Mackie (1982), McIntosh (1998), Schroeder (2009), and Tillich (1963), among many others who argue the opposite.

With respect to the context of the Inexistent God, stories of the bible have been massively challenged. Historical and scientific research, intellectual critique of the bible, archeological discoveries, as well as anthropological research, has analyzed and interpreted writings that contribute to revealing the bible as an unreliable historical document. At best, the critique has enabled these biblical stories to be seen as symbolic language. At worst, the critique exposes the implication that God as real is an illogical stretch of the imagination or a psychological construction of the person's wish-system. And considering wishes and how they contribute to magnificent obsessions, a possible example of such wishing may be seen in the Babylonian Talmud. This Talmud contains 300 arguments for the resurrection of the dead. In an irreverent observation of this army of the resurrection of the dead, some say that if you could produce one good argument you do not need 300, and if you do not have one good argument, then three million won't help. However, when wishes are intense, even 300 arguments do not at all seem unusual.

Rabbi Terry Bard, formerly on staff of the Boston Beth Israel Hospital, sums up the view regarding the Inexistent God, by stating that "Dead is dead," and what lives on are the children and a legacy of good works. At Harvard Divinity School, theologian Gordon Kaufman traced centuries of decline in the concepts of heaven and hell; what is left he loudly proclaims is intellectually empty baggage. Douglas Stuart, an Evangelical theologian, at Gordon-Cornwall Theological Seminary in South Hamilton, Massachusetts, says: "It is a matter of intellectual integrity—the problem is that the mainstream Protestant clergy simply do not believe in the after-life themselves—either the biblical view or any view."

Thus, with respect to the Inexistent God, elements of any God's context have been severely challenged. For liberal Protestants, hell has been fading for a century or more. And with respect to Hell, Universalists, see God as not condemning any-one to such a fate as Hell, while Unitarians generally conclude that humanity itself should not be punished by God in such a way as to consign people to Hell—assuming a God exists in the first place. In addition, in contemporary understanding, it becomes evident that the concept of Hell is much too ridiculous for serious thinking and scholarship. And if that is the case, the question can be asked as to why Heaven remains a serious subject?

With Jews, of course, as indicated by Jacob Neusner (a Judaism scholar at Brown University), Hell can no longer be a tenable idea because life after death is, as he suggests, clearly intellectually embarrassing to modern thinking, and furthermore, to add to this sense of the ridiculous, Jews have been through pogroms, the Iberian Inquisition (preceded and extended in the post Inquisition period with far-reaching persecution against Jews for more than six centuries), and the Holocaust (with six million Jews murdered including one and a half million children), so at least, accord-ing to these atheist proponents, there is no use in talking about Hell, and in the face of such catastrophic circumstances it also could be that there is no use in talking about God.

Again, in analyzing and returning to the context in which people imagine God, in Christian teaching the Gospels are known to be imperfect renditions of what may have happened. Such writing did not gain its final form until the late first century, many years after the death of the assumptive Jesus, and like the Torah (originally claimed as the word of God), it has been decisively identified as being written and completed over a period of 500 years and not at all materialized in a moment.

The Gospel of St. John has been assessed as having been written five generations after the death of Jesus, and in any event, none of the Gospels had not the slightest chance of being written by any of the apostles and instead are thought to be agglom-erational discourses of many authors. And the contradictions are never ending: in some cases sayings attributed to Jesus were perhaps really spoken by James; how Judas really died has been portrayed in ways that completely contradict one another; and, more recently it has been conjectured that Jesus perhaps even choreographed his own betrayal. In addition, the apostles are described in one place as being pres-ent for the crucifixion of Jesus and in another place as fleeing and not witnessing the crucifixion.

As far as the Christian God is concerned, even in the eighteenth century, the scholar Hermann Samuel Reimarus (1778), denied the resurrection and registered other positions considered to be heresies: to wit, that it could be that Jesus was a teacher of morality but that there is and was no such thing as miracles, and that Jesus himself was one of those many apocalyptic visionaries in Roman Palestine. Supporting this position is the nineteenth century literary, theological genius, David Friedrich Strauss (1835), who attributed all of it, to fable. Thus, all of these citations were asserted by those who have pointed to an inexistent God—that is, to a figure who is not a supernatural one.

The Rescue of God

Of course, other writers dispute all the nay-sayers who treat the bible, the gospels, miracles, etc., as literal documents, or as oral reports that are to be taken concretely, or as historically and assumed accepted events. Karen Armstrong's book, *The Case for God*, is one such defense strictly of faith and not one of support for descriptive theological events, and in this respect she stands athwart the atheist/theist disputatious wall shouting: "Stop!"

In her book, Armstrong extols the virtues of myth and faith itself, and decries those so-called religious/theist tracts that consider the manifest content of all theological work as descriptive literal truth. Thus, Armstrong does not spend her intellectual capital in an effort to prove literalism. Instead, she, along with other so-called apophatic searchers (those who consider the deity unknowable) rely more on symbolism, myth, and analogic interpretation. This approach to understanding God is actually reliant on what is felt rather than what is logically constructed. It is almost what is closest to how "spirituality" is discussed. And so Armstrong's approach to God is not static and does not solely refer to any real tangible or static events. Rather, the interface with God necessarily reveals itself as an ongoing process, and in this way, the religious experience is informed by a continued approach to the Divine—almost incrementally, by a series of continuing and successive approximations.

Armstrong, as the representative of the "faith" dimension in the search for God is actually also assisting individuals in the undermining of the process leading to symptoms because in faith are wishes, in a redemptive sense, implicitly promised. Therefore, the illusion of a God (or in the absolute belief that a God exists and understands) offers the supplicant a better chance of prevailing over the disempowerment that accompanies thwarted wishes. A thwarted wish generates anger that is often repressed thus beginning the process that gestates an emotional–psychological symptom. It is in this sense that faith may have ameliorative effects with respect to mental health. Beyond mere empowerment, extreme religionists take it a step further with the idea that disbelief requires punishment. For example, in Islam, modernists question the entire belief that evil ones fall into a pit of hell to be roasted and boiled eternally. In Buddhism, the similarly though more subtle idea of punishment is that after death the individual wanders as a spirit and the common person will return as an animal or slave while an educated person can return in any way he chooses. In addition, Hindus believe in multiple reincarnations as well as special

Hells with special tortures. In countervailing punishment themes, Mormons believe God is married to a heavenly mother who gives birth to spirit children and who then in turn take bodies to inhabit the Earth. Then each person can become a God eventually living hedonistically in pleasurable procreation. And then the spirit children inhabit other planets.

Further, to ameliorate the possibility of punishment, New Age so-called possible avenues of potential curatives include believing in rebirthing, channeling, out-of-body existences, UFO's, phrenology, ESP, telepathy, levitation, and a host of other wish-for happenings. These are beliefs targeted for possible empowerments that presumably help people feel that negative things such as bad luck or punishments can be nullified.

Kathleen McCarthy (1990) reports a survey of a 1990 Gallup Poll indicating that one out of six Americans say they have been in touch with the dead. In addition, one out of ten says they have talked to the Devil, and one out of four reports a telepathic experience. McCarthy further indicates that with respect to at least these reported experiences related to out-of-body claims "… 100 years of parapsychology research has produced absolutely nothing but speculation, stridency, and sham…" Along with such surveys, a Baylor Religion Survey (2007) also reports that 55% of the survey-sample claim they have been protected from harm by a guardian angel, 45% state they felt called by God to do something (*hopefully good*), 23% claim to have witnessed or experienced a miraculous physical healing, and 20% reported hearing the voice of God speaking to them. In addition, Henig (2007) reports that a 2005 Harris Poll showed that six out of ten Americans believe in the Devil and Hell, and seven out of ten believe in angels and life after death, and finally the Baylor University survey reported that 92% of Americans believe in God.

Thus it is clear that large numbers of people persist in challenging the idea of the Inexistent God, and deny scholarly advances in the historical and anthropological deconstruction of the bible. Thus, according to this agglomeration of theist conceptions, the only God that does not exist—is the Inexistent one. And on the growing edge of theist philosophy and scholarship is the issue of faith (as Armstrong would counsel), as the salient issue of man's relation to God. Thus, faith, has been seemingly gaining the esteemed ascendancy in the foundational underpinning of belief in God.

It is thus fairly clear that man's psychology was quite powerful in sharing top billing along with evolution as in the psychoevolutionary discovery of God. In other words, there are psychological as well as evolutionary factors to consider in man's encounter with God. Although psychological pressures are surely infused in the evolutionary process, nevertheless these may be examined as a separate domain—a psychological domain. Some of these factors include transmogrification of the function of organ systems while other variables include psychological mechanisms of the psyche along with spandrels of the thinking brain—all in a ubiquitous search for security, fairness in life, empowerment, and the absolute nullification of disempowerment.

Din Torah (Trial of God)

With respect to the appearance of a spandrel as a function of the thinking brain, in Jewish practice, and from time to time, it is a tradition to have a trial of God, identified in Hebrew as a Din Torah or essentially a check on the works of God (a trial underscored by the laws of the Torah). This Din Torah is based upon God's promise to protect the Jewish people. It is also a check on whether God has been rewarding merit and punishing evil—to reward good deeds and punish bad ones. In Christian teachings, according to Morse (1994), "To believe in God is not to believe everything." Morse states that there is always a call to "faithful disbelief." What it all means is that the call to disbelief is a call for examination of propositions and even uncontested assumptions regarding the entire fabric of the conceptualization of, and belief in God, along with the principles surrounding the entire context of the theology.

In some ways, such a call to disbelief in the Christian search for the valid surround of God, to an intellectual examination of belief in God, and of faith, is akin to the call for a Din Torah as another sort of examination, analysis, or test of God and even help for Him to sustain his Godliness. And as far as merit for good deeds is concerned, Dorothy Martyn (2007), in her book, *Beyond Deserving*, indicates that "… fairness and merit utterly disappear in an in-breaking of a powerful force that transcends 'deserving' altogether." This hopeful theological message suggests that Christianity's fundamental philosophy concerns a utopian vision of the future in which Godliness permeates everything—beyond deserving—and where there is hope that the future can, at any moment, exist as the present—a state of Godly perfection, Godly love. And this essentially not only means the enshrinement of "fairness." It also implies the enshrinement of "goodness."

In contrast, the Din Torah of Jewish tradition may be more focused on controlling the unruly evolutionary state of current human existence—the absolute present—where impulses and needs flood the psyche so that people are essentially gyroscopically imbalanced. Being imbalanced means that people are psychologically challenged, requiring all sorts of ethical and moral policing—including the presence and influence of philosophers, humanities professors, psychologists, clerics of all stripes, and even police. In the context of the Din Torah, there is really little hope for any sudden miraculous transformation of the quite imperfect and unfair present into the perfect Godly future.

With respect to the Din Torah, conclusions of the four types of Gods (Intervening, Impartial, Irrelevant, and Inexistent) lead to the following verdicts based upon implications of their behavior. However, it needs to be noted that authors such as Karen Armstrong, in her book, *The Case for God*, clearly makes the case that all of this theorizing about God's intervention or impartiality, or irrelevance, or even inexistence, emerges (as Freud would agree) only from the thinking known as logos—the cognitive attribute of "reason" (the word). Yet, Armstrong cites the issue of "mythos" as something closer to spirit, also implying that the search for God resists logos, is beyond logos, and is a never ending search for meaning that harkens to an encounter with the mythos of the ages—implying that the presumed reality of one's soul is a microcosm, that which recapitulates the mythos of the ages—the search for God.

The Din Torah (Trial of God) is based in "logos" with full awareness of how its verdicts might be different if the tradition of the Trial was based rather in "mythos." Of course the pivotal motive in the creation of the Din Torah concerns the issue of *fairness*. In other words, and simply: Has God been fair?

Implications of the Trial of the Intervening God

The Intervening God to whom one prays and asks for something—a wish, a hope—behaves essentially in what could be considered an immature manner. The immaturity is seen in the inability to distinguish between important and unimportant prayers, sometimes answering unimportant ones while neglecting important ones. For example, favoring the wife's request for a particular color that she wished her husband would select for the automobile he was purchasing but not acknowledging the prayer of a mother with a sick child is seen as an absence of understanding by the Intervening God. This sort of absence of understanding resembles the behavior of a very young child who cannot appreciate the difference of important from unimportant issues. Such immaturity was also characterized as reflecting an incompleteness akin to the one postulated by Alfred North Whitehead (1979), in his "process theology." This same idea of the incomplete God (here associated with immaturity) is also theorized by many Jewish theologians, and by many Christologists—those who believe in Jesus, the man, and not Jesus, the supernatural God.

Gerald Schroeder (2009) arrives at a similar point but rather than considering such a God to be immature, sees this intervening God as someone who is still learning how to create, and continuing to refine his creation with changes.

Nevertheless, whether the God is immature or simply learning on the job, such a God might benefit by consulting people, perhaps to be helped to mature and/or to acquire higher education. Thus, the act of deferring to this Intervening God seems of course, to be ridiculous—despite Eagleton's (2009) reminder that scripture addresses all the existential questions such as love, loss, happiness, guilt, agony suggesting that these make up for God's irregularities.

At the most extreme, such a God could also be judged guilty of capital crimes against humanity; that is, this God is an intervening one but did not intervene to prevent genocides, torture, slavery, and so forth. However, even as a result of the highest order of theological purity, capital punishment would be anathema (as it would be under secular atheistic principles), and therefore, the Intervening God's death sentence would need to be commuted, especially in light of this God's immature stature.

Thus, one possible facet of the nature of God includes the characteristic of immaturity rendering Him less able to judge issues of fairness.

Implications of the Trial of the Impartial God (Non-intervening)

The Din Torah judges such an Impartial God to be immoral; that is, the Impartial God's promise to self, to not under any circumstances intervene in human affairs, was seemingly more important than reversing that decision in the face of human horrors in order to prevent catastrophic events of all sorts—and to do so throughout history! Here too, this Impartial God is judged guilty of capital crimes. Yet, clemency is granted because it is felt that this God was psychologically paralyzed, and because of such mental disability, rendered in essence catatonic, stock-still, and frozen.

In the face of profound cruelty and an injustice pervading the world, this Impartial God was for all intents and purposes, immobile. Further, it was too much for this God to understand and to recognize that the world was much different than had been anticipated and that given this enormous evil in the world—this leaching miasma penetrating the very fabric of the world—the promise to remain impartial would, to say the very least, need to be changed in order to plainly stop the carnage.

But this Impartial God apparently could not alter the promise to the self. Therefore, such a God was pronounced incapacitated and given an insanity plea. The treatment for such a pronouncement was to provide daily psychotherapy for this kind of God until such time that the catatonia retreats and the God is able to become less God and more person—a person concerned with fairness.

Thus, another possible facet of the nature of God suggests that such a God either may be morally challenged, merely inert, or even seriously psychologically incapacitated.

Implications of the Trial of the Irrelevant God

The Irrelevant God is the remote creator of worlds and correspondingly has worshippers who are entirely interested in allaying personal anxieties—not at all interested in larger human affairs. This Irrelevant God is likewise not concerned with large events such as genocides. Rather, such a God seems more intrigued by events entirely unrelated to individuals and similarly, individuals who are connected to such a God are only focused on idiosyncratic personal tensions and not at all on a defined or specified God. Therefore, this God is probably never prayerfully invoked for the good of mankind or for other altruistic concerns. Instead, this God is called upon in the hope of supporting reduction of anxiety and therefore it is an unconscious practice by such worshippers to assume and naturally expect that the Irrelevant God to be actually an Intervening one.

The Irrelevant God here discussed is equivalent to the equally primitive notions in earlier ages of Rain Gods, and Sun Gods that were seen as intervening and always powerful enough to induce supplication for the purpose of reducing anxiety over any number of survival concerns. The Irrelevant God, therefore, is most likely the equivalent of an Anxiety God (only on a part-time basis), with no universal meaning whatsoever, and we might say that this Irrelevant God may really be us—that is to say that our anxieties lead us to such a God and therefore this God cannot be convicted simply because we will not convict ourselves.

Thus, another facet of the nature of God concerns the person's denial of the irrelevance of a God that simply does not care, and in its place the projection of caring—the implicit hope that such an irrelevant God will be persuaded to become relevant by virtue of such implacable worship.

Implications of the Trial of the Inexistent God

In order to examine this Inexistent God, it is helpful to refer to the idea of covenants. In the Old Testament it is the covenant that generates a chosen people. In the New Testament, it is a new covenant and perhaps a new chosen people. Thus, everyone wants to be chosen; for example, "China" means center of the universe, and in America, "Manifest Destiny" has a similar meaning.

However, when discussing the possibility of the Inexistent God, there is a somewhat different perspective on the idea of "the chosen." If it is an Inexistent God, then rather than a focus on who is the chosen, it becomes more useful to think about "choosing." And "choosing" implies that people can choose to be ethical and moral rather than unethical and immoral although a person's early experiences may make it virtually impossible to do what one knows is right, or even to know what is right. Yet, people can think and feel and know what is right and what is wrong. In this sense, it seems clear that everyone is basically chosen insofar as evolution favors cooperation, altruism, and service to others as highly adaptational. It is tempting to also say that as an absolute (even outside the theological or evolutionary rationale for understanding good from bad), to choose to do good seems to be the core revelation, at least in the positive "upside" aspect of the psychology of empathy.

The question becomes: Who tells us that to choose to do good is naturally good? Is there an authority, implicit or otherwise that influences us in the ways of moral choice? And in terms of the Din Torah, is there a proclamation there that can offer an answer to this question of whether there is some source possibly that helps us know what is right?

Thus, another facet of God's nature (in the inexistent sense) concerns the implication that such a declaration of God's presumed inexistence requires man to decide right from wrong without any Godly help. Such an ability (opportunity) to decide right from wrong will largely depend on the appropriateness of a person's development whereby epigenetic considerations were operationally successful; the child in its development was treated with consideration, love and appropriate responses. Hence empathy, that great teacher of moral and ethical instincts and inclinations becomes, in a manner of speaking, one's internal God. And this empathetic internal God is another perspective related to the "psycho" of the psychoevolutionary approach to the understanding of the concept of God. This issue of man's capacity for empathy is by no means meant to be an argument for the replacement of God or a refutation of God. It is simply one of several constituent factors in the gestating cauldron of the psyche, of the psychology of man that ultimately has a bearing on moral and ethical behavior.

As a postscript to the analysis of the Din Torah and the variations of the possible types of "the" God, we can also now wonder about an existent God's intention with respect to the concept of the *psychoevolution of God*. It has been assumed earlier that if God in fact does exist, then such a God permitted evolution to occur in order for the ultimate appearance of Homo sapiens characterized with an advanced thinking brain that can create consciousness so that the person correspondingly can understand how to find God, and then to actually achieve this Godly discovery. The focus on this evolutionary unfolding is strictly related to biological transformation. Such transformation can be clearly seen by and within anatomical configurations as well as in the behavior of animals throughout the phylogenetic scale; it remains as the trajectory toward the development of the thinking brain. In the same way, the gradual appearance of cortical development and the synthesis required for the mind and psyche to develop represents the "psycho" (psychological) of "psychoevolutionary."

Given the actual nature of the world and of its inhabitants, a predatory/prey template seems to be an outgrowth of all of this development—biological as well as psychological—so that the unfairness pervading everyday life, and the pain and suffering of people, implies that a Din Torah can be quite necessary. It represents man's moral ascendancy in a way that prevails with God, implies that man does not fear God, and gives greater credence to the need for holding a possible extant God accountable for this unfair, unfeeling, predator/prey earthly template that in turn gives the Darwinian "survival of the fittest" program a Godly okay as preferred over the Thoreauan program that states it is fitting for all to survive. Here, Thoreau is challenging the idea that one must live only if another dies. In this sense what is challenged concerns the connection of predation with that of survival of the fittest.

Perhaps in order to rebalance this imbalance of Darwin over Thoreau in relation to biblical texts—an implicit sequential imbalance in evolutionary history—it should be noted that biblical texts were developed to locate the discussion away from this imbalance and toward an overall and overarching concept of the basically good God—ironically, the one we cannot understand. Such biblical and religious/theistic orientations bring the faithful as well as doubters to perhaps believe that the evil in the world may reflect something we do not know about God's plan and that such evil create other syntheses that lead to greater glories—and in addition, these biblical teachings actually seem to excuse all this evil by implying that the victims will be rewarded in the afterlife while the victimizers will burn in Hell. In the sense of seeing biblical "knowledge" as hypnotic manipulation, many have considered deceptions and myths of the bible to validate such manipulation.

The Proclamation of Din Torah

Lloyd Graham (1975), in his treatise on deceptions and myths of the bible puts it this way: "There is nothing holy about the bible, nor is it the word of God. It was not written by God but by power seeking priests." Graham calls it "a priest perverted cosmology." And with respect to the Din Torah, it should be noted that it was the Jews during the Maccabean period—about 160 or 170 years before Christ—who questioned the justice of a God who demands obedience in life but delivers the wicked and the faithful alike to oblivion. Of course, the answer was resurrection. The just would be united with God; the evil one's relegated to Hell. And in Islam a similar scenario exists in which after death two angels question you so that when the end of the world arrives and Allah declares final judgment, either you go up and enjoy everlasting life or down into Hell.

For Hindus, reincarnations can take thousands of years and then if one is fortunate enough to end up in Heaven that one will be blessed with music, flowers, choice foods, and beautiful women. Of course this omits considering what women would want in Heaven unless of course, where pleasures are concerned, Heaven is a place arranged differently for women, or not for women, or that women simply do not count and therefore won't ever get to Heaven. But if per chance they do count, what about women who are not considered beautiful—and who decides? Are these not so beautiful women consigned to another place?

In sum, Neusner, the Brown University scholar has also stated that "the idea of a life after death is clearly an embarrassment to modern thinking; most major philosophers have ridiculed it, but it is just as clearly [*still*] the touchstone of all religion." And yet, many of the most respected literary people (although certainly not all, and perhaps not the majority) agree with Neusner in the belief of the Inexistent God. Such personages (in this case from one historical period) include F. Scott Fitzgerald, Beckett, Camus, Hemingway, Ionesco, and Jean Paul Sartre, to name a few.

The Din Torah then proclaims that the Inexistent God is actually the political power existing at any given time. And whatever this political power wants is what will be. If that power wants the populace to believe in any of the Gods discussed here (Intervening, Impartial, Irrelevant, Inexistent) that is what will be. This sort of "power" can, and frequently does become tyrannical, and therefore obviously needs to be restrained in the context of actual democratic checks and balances.

The essence of the Din Torah is a call for debate and conversation—speaking to power. It is the psychoevolutionary epigenetic revelation of God knowing right from wrong and consciously knowing and supporting nonexploitative, good things unto others so that no matter the particular ideology, truth will always trump ideology and the personality of any God, will never be as important as the actual practice of decency in human relationships.

Coda

The story goes that God instructed Moses at Mt. Sinai to divide the authority over the people into the roles of Rabbis and prophets. The task of the Rabbis was to teach the people the rituals of the religion. The task of the prophets was to help the people navigate their secular lives. The problem developed for prophets because whenever they witnessed governments exploiting the people they encouraged rebellion. Of course, because of such so-called subversive practice, the prophets were suppressed, and eventually disappeared—they went underground. It was only centuries later that these prophets gradually emerged, reappeared, but in a different form, disguised as it were as philosophers, humanities professors, historians, psychoanalysts, and other such presumed truth-seekers. And many of these prophet-equivalents became more interested in secular life and less devoted to issues of supernatural interest.

Thus, the new Din Torah becomes not so much a trial of God (with its implied blame psychology), but a new direction in which the goal is concern over how society can be encouraged to enlarge the moral context by focusing on living a just life, and to respecting the inviolability of others.

In the following chapter, the ideologies of certain groups will be examined with an eye toward showing how a blame-psychology interfaces with the various motives of religious and other groups. In considering the psychoevolutionary perspective in the search for God, it seems necessary to examine the *emotional* component of the issue of psychological mechanisms that people implement in order to find greater security and sense of balance in their lives. However, the question becomes whether the need for assimilating oneself in a given group is realistically and actually sustainable (or even good)—especially in the context of large sociological arenas composed of many different kinds, and motives, of groups.

It boils down to this: Is my God better or "realer" than yours? And by logical extension, the notion that "we" need to destroy "you" because "we" are better than "you." This is what people actually do when the personal psychological relationship with a God is translated into sociological practice. Is it possible that under such conditions of presumed superiority, the internal need for Godly affiliation can become externalized toward other groups in the form of a "blame-psychology?"

Such a possibility raises the issue as to whether the psychological component of the psychoevolutionary perspective of God, like a spandrel begins to serve other masters—for better or worse, these surrogate God-heads or what is known as cult leaders. In addition, an examination of the infrastructure of groups and their core motives can help in the understanding of the psychoevolutionary perspective of God—beginning with the function of the animal's tail along with the bio/evolutionary phenomenon of animal group behavior, becoming then the precursor to the appearance and function ultimately of Homo sapiens advanced thinking brain; all of it presumably driven by a biologically fueled adaptational imperative for safety, security, peace of mind, the need for fairness, and the wish for empowerment along with an absolute nullification of disempowerment.

This discussion will begin in the following chapter with how groups are organized so that they can gain advantage largely with respect to owning or possessing resources. In the sense of human groups such resources are people. We begin in human groups with the notion of the connection between a "blame-psychology," and its putative relation to God.

Chapter 4
God, Group, and Blame Psychology

Man's either real or imagined encounter with God also, and very frequently, relates to man's corresponding functioning in, and basic attraction to groups. How groups integrate individuals into their prevailing norms and ideology will now be examined. It is basically an analysis of the group mind, as an equivalent to the so-called group mind of animals in which individuals of the group through evolution have an enhanced ability to survive, be safe, and therefore to be more secure as a direct result of group membership.

This evolutionary path that in ever gradual steps (along with mutational moments), selects for anatomical changes and transmutational or transmogrified functions—within newer and newer life forms—is easily exemplified in seeing how the animal's tail functions in various ways that include security and alarm signals (as the tail can swish or wag to help protect the rear from attack) similar to the security function for individuals in groups, and then again similar to the security function inherent in the development of advanced thinking—the human thinking mechanism—the cerebral cortex or man's thinking brain—the mind; it is the brain/mind that can imagine, can find, can encounter God or enable God to encounter man, or even to not believe in any God whatsoever.

In part, the analysis will reveal the basic elements of the infrastructure of groups in order to see how group forces operate to affect individuals. Ultimately we are interested in understanding how the individual's relation to the group can affect that person's encounter with God or belief in God, or that person's quest to confirm and profoundly reinforce a Godly affiliation.

The central tenet of this examination can be labeled the uncovering of the punitive mechanism of groups. In essence it can be expressed as "my group is better than yours!" Essentially it becomes the need to blame (to identify wrongdoing), and to punish, to be punitive. At the basic level of intensity it is the person's attempt to satisfy the longing for fairness. On a middle intensity level the punitive impulse reinforces a sense of superiority of one's belief system, and on the highest intensity level it is a sense that retribution is required for whatever is deemed to be wrong with others. It is the person's (society's) justice-oriented attempt to rectify, which in some cases represents in negative ways, a satisfaction of righteous indignation.

H. Kellerman, *The Discovery of God: A Psychoevolutionary Perspective*,
SpringerBriefs in Psychology, DOI 10.1007/978-1-4614-4364-3_4, © The Author 2013

This sort of mentality requires reward for merit and punishment for wrongdoing. Of course, as shall be discussed, the assessment of merit as well as the attribution of wrongdoing depends largely on the particular perception of what constitutes right and wrong and such assessment in turn depends on the socio/psycho/religio/political orientation of the assessor.

Fundamentally, behavior designed to satisfy righteous indignation serves the purpose of empowering one while disempowering the other. Another way of seeing it is to start at the high intensity level of the punitive impulse in which the need to punish is the person's attempt to actually satisfy a variety of needs—from frustration, the spewing of anger, and need for retribution toward others straight to the further end, that of longing for fairness and justice.

In general when one conquers another as in some way doing them in, or when one seeks purity by identifying others who are ostensibly impure, then the issue of blame psychology becomes quite relevant to understanding individual as well as group behavior. Such a dynamic can be seen in the ugly history of scapegoating. In addition and directly related to the theme of this book is the phenomenon of a devout person, who coexists with the ritualistic empowering worship of God on the one hand, and on the other, psychically engages in personal fractured moral behavior. Of course in animal groups fighting within the group and the fighting of group against group reveals motives that are not as seemingly complicated as motives found in human groups. Nevertheless, the sense of ownership, conquest, scapegoating, self-interest, and security remain quite equivalent both in animal and in human groups. Most of these characteristics in human groups will be considered in the following sections on the analysis of the group's blame psychology.

Frequently, a blame psychology is essentially the condition of faith-based belief (whether theist or atheist) and becomes fanatical (or leading to fanaticism) as in caste systems of scapegoating or in racial discriminations, as well as in other forms of the quest to be superior—including those of political ideologies. In addition, the need for superiority can, in a practical sense, generate behavior on the basis of palpable hysteria activated by the excessive unity, chauvinistic nationalism, and cohesive togetherness (especially in the form of high cohesion), that then is enthusiastically embraced and expressed by a group. Wilson (2012) examines this issue of superiority in his analysis of group function in evolutionary development.

Generally, the need for validation of superiority, or of the need to blame, or, in an overall sense of the need to join with others in order to achieve an inner sense of security (by identifying with like-minded people), or the belief that others need to be brought to some elevated frame of thinking, is frequently found in the affiliation with a group characterized by a specific ideological identity. This focus on "my belief is the true one while yours is not" is ubiquitously seen in religious and political groups. It is certainly true that voluntary affiliation with such a group that resonates with one's needs in a congruent way, can offer exquisite peace of mind and a sense of safety. Further, reassurance of one's needs is also gained in a group affiliation where the search for mentors, icons, idols, or simply people to admire, are available. In most cases it becomes a search for the all-protective, everlasting powerful parent—in the discussion herein related to the abstract God.

And this is the essential spandrel—psychological mechanisms that originally created peace of mind by encountering God, but such mechanisms can also be operationalized by turning affiliational group sensibility against others. Therefore as an apparent unintended evolutionary side effect (the spandrel), gaining such peace of mind takes the form of: "We are superior to them." Joining a group that may validate one's "rightness" by asserting an agenda for conquest then suffices as an assurance regarding the appearance of such a self-assessed superior mindset—whether conscious or even vaguely assumed.

As an example, in the affiliation with a religious group, the search for a clearer path to God is enabled, and the person's particular orientation toward devotion can be validated on the one hand by other group members who share the same feelings, and on the other by the effects of the group-as-a-whole—that is to say by the overall cumulative agreement force of the group. And the implied point can be expressed as a question: Where does this agreement force of the group get directed and toward whom?

To begin this analysis it could be useful to attempt to understand "affiliation" and its purpose. When one joins a group on the basis of identifying with its characteristic spirit, the ethos of the group, and in order to feel aligned with its basic ideals and purpose, then the issue of the cohesion of the group and its affiliational power becomes important to understand. Understanding this affiliational power of the group is also a way to understand the nature of how a blame psychology can be an instrument of the group's deepest cohesion. And the fact is that high cohesion can be either wonderful or horrible with respect to the aims of a group. Historically, and with respect to religious and political groups such cohesion-power of the group has led to horrendous outcomes.

In the cohesive group, when the sense of blaming others is evidenced by the attitude and behavior of individuals functioning toward these "others," it is often as an implicit mandate inherent in the group's ideology. Such blaming behavior is also often, and by definition, assumed, certainly in cohesive religious groups to be approved (blessed) by God.

The Cohesive Group

The issue of group cohesion and its ability to align and absorb membership agreement with respect to the goals of the group as well an analysis of the psychosocio composition of such cohesive groups has been examined by a number of authors (Kellerman, 1979, 1981; Lifton, 1979; Lott & Lott, 1965; Redl, 1942). Generally, all authors conclude that the aims of the group within a strong cohesive structure can generate singular purpose in the membership to do whatever it takes in order to accomplish group goals. And such zeal when directed toward others usually produces results that galvanize cooperation within the membership. When such a group becomes a punitive one (blame-oriented) toward identifiable others, then any action toward those "others" can, and most often will, cause destructive outcomes—and usually of course, with an underlying, permeating, and punitive sense of righteousness. As stated earlier, such is the historical record of religious wars as well as attendant political strife.

In this respect, the cohesion of groups can be examined on the basis of what could be termed the specific deep punitive structures that can exist within any group's evolving or existing culture—including, but not solely limited to religious cultures. And the essence of a punitive group culture leads to the creation of a blame psychology with its ultimate justification that occurs either by invoking God's will and rightness or by some other adoration of an equivalent God-like ideological figure who is idealized and/or feared.

As the God–idolatry that characterizes most groups becomes heightened, so too does the cohesion of the group correspondingly increase. This idolatry increases even to the point beyond cohesion—that of adhesion, the net effect of which is to affirm and validate the superiority of a particular God—or leader. Essentially it is the psychological/emotional institutionalization of God's presence (symbolized by the presence of the group's leader) that screams the slogan: "God is great, God is real, and he is ours!" And it is the "ours" (the group) that becomes the validation of one's belief system, further reinforcing the accumulation of cohesion.

The properties of such a high-cohesion (adhesion) group is thought to be composed of affiliational superego needs of members; that is to say, the sense of which behaviors deserve reward, or which punishment, or condemnation. In addition, leadership contributions to the superego alignment of the group (what is considered good or bad) significantly influence general group performance. And finally, the nature of the group's development and the evolution of its particular myths and legends further support the group's distinct identity. Thus, in summary, such groups seek Gods, or charismatic leaders to be "Godheads" and such leaders support the group ethos regarding what is good and what is bad; what needs to be rewarded, and through blame, what needs to be punished.

Derailment of Man's Encounter with God

With this analysis of the nature of group culture and its belief traditions it becomes possible to see that cultural imperatives implicit in gradients of group cohesion (along with constituent blame psychology) become infused into belief systems of individuals; that is, influencing individual thinking and behavior. This infusion of such belief also bleeds into the component of psychological dynamics to the point of contaminating the "psyche" component and focusing the "psyche" component on issues other than personal autonomy and productive positive pursuits. It is as though the thinking brain is regressively transmuted and infused into group behavior so that rather than an encounter with God as the goal of the psychoevolutionary thinking process leading to God, the group now can be as stated, regressive, and its Godly "seeking-safety" mooring loosened; that is, the group under such co-opting pressure can become an agent of maladaptive intent.

In view of the above, general agreement among researchers of the psychology as well as the sociology of cohesive groups indicate that peak cohesion, although certainly efficient in the pursuit of group goals, can be potentially dangerous with respect to intergroup relations, as well as to individual freedom of expression. Thus in highly cohesive groups, individual members can be rather easily co-opted by the group's leadership and overall group influence to display approved behavior. These behaviors are also implicitly reinforced by the very identifiable nature, the spirit and general ethos of the group. Religious extremism as well as its kindred spirit, political chauvinism are examples.

What this group force inexorably means is that in highly cohesive groups, the independence of individual members can be compromised or even entirely usurped by the overall dictates of the group's core ideology. Such a group can be a tyrannical force that essentially confines any member in an emotionally and psychologically internal (psychically interior) imprisonment. This restriction of internal freedom (seemingly voluntary) is also in many cases a usurpation of individual will preventing independent thinking and behavior.

The usurpation of individual will in highly cohesive groups occurs because in a general sense, the nature of any group is reflected in its core motivation; that is, the group wants to reinforce its own structure or its own reason for being. And of course, what makes it all work is that the group membership is then driven to support such affiliational pressures.

Most newly formed groups will generally seek to reinforce their structures. For example, classroom groups are able to construct procedures for the exchange of information in order to facilitate the learning process. Such a developed structure encourages better class performance as well as increased class participation. Political groups strive for ever increasing membership by appealing to a common set of principles and/or interests. Religious groups seek converts or worshippers and/or project a sense of sacred affiliation. Psychotherapy groups try to establish rules that tend to stabilize the group so that members can work toward attaining a greater sense of interior psychological freedom. In each case, the structures and ideology of the group are designed to create adaptational opportunities for members that contribute to overall development as well as to consolidate a cohesive environment for the purpose of achieving goals, or in some cases to simply express its personality and implicit ideology.

The inclination toward cohesion is the group's own survival instinct to persist indefinitely. It is a basic life-force of the group, a group tropism—that is, a singular imperative for the group to become self-sustaining and everlasting. In a sense, the group can become a God equivalent, an everlasting social context in which membership assures protection and safety.

In contrast, when the underlying structure of the group is not adhered to (in most cases due to a weak leader—a poor example of a strong Godhead), then cohesion is compromised. However, in those groups where cohesion is highly developed, an adherence to the group ethos can become so powerful as to have the effect of producing not only a rigorous adherence of membership but rather a rigorous adhesion of members in which practically all autonomy is forsaken. When relinquishment of autonomy in individual members occurs then the possibility of a figurehead, equivalent to a strong Godhead, can set the tone of a blame psychology as an essence of the group culture. When such a group culture has been synthesized and functions accordingly then aiming it with malice toward any other group can be a devastating encounter for that other group. History demonstrates that such a phenomenon regarding the malice of highly organized and cohesive groups acting aggressively toward the "other" has been typical of religious and/or political ideological strife, and at the extreme, in actual wars fueled by dire hatreds also leading frequently and even inexorably to genocides. Lifton (1977, 1981) discusses this point extensively in an analysis of victimization. This is what is meant by the derailment of man's encounter with God; that is to say, socio/psychological agendas in highly cohesive groups with malicious motives can generate and even accelerate a momentum with respect to untoward goals. In religious groups malevolent behavior in the name of God will triumph over Godliness. Here, survival in the form of triumph over others is the goal rather than fairness seeking.

The Relinquishment of One's Ego to the Group

The basic definition of cohesion refers to the existence of an assumed group attitude with respect to its own reason for being that in turn is accepted by its membership as a context and framework for affiliation. It is when the group's system of attitudes and values become crystallized, that it can be said to have cohered. In this sense, the culture of the group can generate intense motivation in individuals to be part of the group and to help keep the group on course toward the achievement of particular goals, and, in addition, such cohesion also reinforces and even acts to continue to fortify the alignment of group attitudes.

Bion (1959) even proposes a startling theory that postulates that the group is able to transform individual member attitudes into unified group emotions and that such transformation of individual attitudes into unified group emotions reflect deep unconscious infrastructural elements of the group. In essence, such deep infrastructural elements (the group's underpinnings) can be understood as punitive structures that largely determine the personality of the group, its collective ethos, and its basic attitude toward other groups. In fact, that the group can have punitive license also accrues to its similarity to the prerogatives associated with the power assumed by Gods especially in targeting other groups with a distinct attitude of righteous indignation.

This entire process of fortifying the cohesion of a group and reinforcing its ideology is a major factor in the development of religious sects as well as national political movements—all intended to validate belief.

In terms of the psychoevolutionary discovery of God, the affiliation to a God (or group) and its value in providing individuals with a sense of power can be seen as profoundly persuasive—even to the point of sacrificing one's life. It is a relinquishment of one's ego to the power of the group for instance as in the thousands of extremists who are able to co-opt participants to carrying out suicide bombings.

Such suicide participants would not agree with the interpretation that they were co-opted—demonstrating what is meant by relinquishing one's ego to the group, or similarly, to the God.

In the deepest, most intense affiliative commitment to the God (or group), the need for evidence of God's superior position frequently requires a punitive attitude toward other competing God's (or groups). In the action of highly cohered groups with such deep commitments and beliefs, punitive actions are based (and gain their rationale) from the deep punitive structure inherent in the group's ideology (religious or otherwise).

Therefore, how individuals treat the issue of a Godly affiliation is frequently to try to determine how to increase a sense of security by joining with like-minded people in order to affirm superiority over others—"God is on our side!" It is the person's basic identification with God.

Thus, identification with God is the personal instant that reflects a cognitive/emotional crystallization of the recognition and conviction that the God is there, that God exists. This instant or personal moment that one intuits idea of God (coinciding perhaps in a sudden affiliational epiphanic revelation), is the birth of the idea of "origin" or "discovery" in the phenomenon defined as *the discovery of God* from a *psychoevolutionary perspective*. It is the moment when the thinking brain identifies either an object (God) or engages in an unconscious process of psychological projections that can create such an identification, leading in turn to at least a temporary state of emotional balance, a relieving sense of safety and security, and consequently the achievement of a sought after emotional state in which peace of mind and personal empowerment prevails, also ensuring the neutralization of disempowerment.

Punitive Structures of Groups

As a reminder in this discussion of the nature of cohesive groups, we are interested in uncovering the relationship between issues related to God and how such issues involve individual belief systems. A key element of affiliation to a group involves one's needs and one's impulse to express whatever frustration (and therefore anger) that exists. When such needs for expurgation are placed into a theological context the rationalization and extrusion of personal hatreds are rendered acceptable by one's actual group affiliation: "You see, it's not just I that feels this way, it's everyone that feels this way" (referring for instance to hatred, frustration, hostility and righteous indignation, and the "everyone" meaning members of the affiliated group).

In this respect, any person's affiliational needs are in part characterized by the punitive nature of their own psyche as it corresponds to the inherent punitive structural ethos of the group. In other words, the most powerful affiliative element attracting any member to a group will be a function of the extent to which the group's punitive structure corresponds to the punitive or superego structure (critical agency) in the personality of that particular member, and reinforces the idea that "God is on our side." In other words, punitive structures are characterological wellsprings upon which group attitudes are based. These punitive structures are also orientations toward intergroup relations and are defined as the typical ways of managing individual tensions coalesced into group tensions—tensions that are confronted (as in a therapy group), expiated or suppressed/repressed (as perhaps in a religious group), or projected (as in all sorts of groups that direct hatred toward others in order to keep one's own group pure and without criticism). The given, that "God is on our side," is then the prevailing mantra of such groups.

Another way of thinking about these deep affiliative superego structures is to ask: How is blame being managed? Is it confronted at the source, is it suppressed or repressed, or is it expiated, or even projected? The punitive structure of a group (akin to the group's superego) may be compared to the superego structure of individuals, and the psychoanalytic understanding of "superego" can be approached and deciphered when examining the psychology of blame—true for individuals, true for groups. The principle is The more intense and serious the blame, the more intense and severe the superego.

Types of Punitive Group Structures

In characterizing the nature of the so-called superego structures of groups, three basic types are revealed. The structures of such groups can be identified as the *Extrapunitive Group Structure, the Intropunitive Group Structure,* and the *Impunitive Group Structure*. Each of these structures contains a gradient of intensity or particular potential to activate one's loyalty and devotion to a leader or Godhead. This formulation of punitive structures was originally presented by Murray, Erikson, and White (1938) and discussed also by Foulds, Caine, and Crasey (1960).

These are basic modes of behavior by which the group manages anxieties, conflicts, and overall tensions, and in addition, offer members a sense of safety and security. The inherent power of such relative punitive structures contributes to the affiliational identification with the group and helps convert individual affiliational identification into the accumulated affiliational cohesion-energy of the group as a whole. Thus, a person's tendency to want to blame or punish becomes a justified impulse because its nature and management will be validated by the ideology of the group to which one is affiliated—that is to say, the individual now has the opportunity to align oneself with the group and in an approved fashion.

The three types of cohesive group patterns that manage tensions and typical impulses, do so, either as stated, through confrontational inclinations, expiational needs, suppression/repression mechanisms, or in a straightforward projection of such impulses onto others.

The search for a Godhead arises out of one's need system, and this Godhead that fuels the loyalty to the group then correspondingly becomes a surrogate object of permission to express individual feelings (hatreds, hostilities, and frustrations that coalesced into the sense of righteous indignation). This psychological surrogation is the "psycho" of the psychoevolutionary viewpoint regarding the thinking discovery of God.

Interestingly, this search for Godheads seems to be a recapitulation, a repetitive, continual reenactment of the psychoevolutionary march to the protective God, as the abstract parent whereby from remote historical time to the development of a thinking brain, evolution enabled arrival at a point where God could be conceptualized, recognized, and believed.

The Extrapunitive Group Structure

Extrapunitive groups tend to search for scapegoating avenues of problem solving. This kind of group cohesive culture was exemplified in the development and activities of the Nazi movement. All difficulties and problems of the mainstream group were cast in a searing and ultimately devastating victimization, mostly of Jews, and also including other targeted groups. The intensely high cohesion of this Nazi became bestial when it was fueled and directed by an excessively intense cohesive extrapunitive group culture with an exceedingly idolatrous personification of its leader. The idealization of Adolph Hitler was a study in itself in the psychology of allegiance, worship, and maximum idolatry. It was, without a doubt, a passionate enshrinement of a God or Godhead constantly expressing righteous indignation.

An analysis of the extrapunitive culture of groups is also a useful heuristic (as implied by Lifton, 1979), for understanding subjective, so-called "superior" morality needs of the mainstream victimizer group. Taken by themselves, individual leanings toward punitive behavior can be understood as "immorality" impulses, and even immortality needs. Within the group's infrastructural instinct such immorality impulses along with immortality needs become the group's signature need to persist. Therefore, the group infrastructural instinct entails a continued need to find targets. Such a conclusion is also made by Norenzayan (2006), who refers to this sort of extrapunitive manifestation as "cultural phenomena."

In addition, victimizer groups are highly dangerous toward their own members as well as their targeted scapegoats as the victimizers will know who you are, and of course, where you are!

The Intropunitive Group Structure

Intropunitive groups tend to look inward. Frequently, rather than analyzing one's personal behavior and moral and ethical responsibilities, members of such groups become concerned with personal sin and the expiation of that sin. Many religious organizations are examples of intropunitive cultures.

However, such cultures also can be dangerous to others because in the hands of an ascribed greater force, the location of sin can be seen in the hostile or otherwise dangerous, impure motives of those outside the group. The fighting of religious wars or the blessings of God bestowed on the soldiers of opposing armies by their respective chaplains are examples of how intropunitive cultures may from time to time resemble extrapunitive ones.

Cultism in the guise of religion also generates great cohesion and yet such extremist cultures can produce the potentially dangerous to the unimaginably homicidal in which blame can lead to genocide, or even to the suicide of all the individuals of the group with God-like cult figures leading the way—perhaps even toward mass suicidal/homicidal acts (the Jonestown massacre), and even toward purely mass homicidal acts (the Holocaust, the Turkish genocide against Armenians, the killing fields of Cambodia, genocide in Sudan).

The Impunitive Group Structure

Impunitive groups are those in which members tend to examine their own motives, feelings, and learning processes. These groups are not culturally guided by scapegoating or victimizing needs, and they will also not usually insist that members unconditionally follow collective instructions. In this sense they are not inherently hostile or enraged toward members who disagree or toward other groups. They are not implicitly paranoid and any blame psychology of the group remains quite uncrystallized and quite irrelevant. Examples of groups in which the culture may be an impunitive one include psychotherapy groups and educational groups.

In the impunitive group such as the psychodynamically oriented therapy group, intergroup relations can be egalitarian because tensions and anxieties of individuals must be confronted and perhaps reworked. This is true also in educational groups with no particular or ideological agenda other than an educational one. Thus, there is no ultimate motive generated by the impunitive group culture to expiate, displace, or project blame, and presumably there is no particular, focused, or even covert desire to elevate the leader to God-like status. Respect, yes. Love, yes. God-like status, no.

In contrast, and with respect to the punitive structures of group culture, such punitive cultures will always determine group attitudes and profoundly influence intergroup relations. For example, in the extrapunitive group, intergroup relations are inherently hierarchical and antagonistic and governed by an authoritarian structure. It is similar for intropunitive groups because followers depend upon the authority hierarchy to issue expiation, typical codes of behavior, and rather strict obedience to imposed symbols. This often gives rise to the persecutory idea that out-groups need to be controlled or ostracized. At its most neutral, it suggests that the out-group should be ignored, perhaps reeducated. In this sense, the idea is that the out-group should be pandered to, or at times actually deferred to in part so that the affiliated punitive group can continue to see itself as good and desirable, and to insist on its assumed and self-proclaimed inherent rightness. In all, such groups in their euphoria as "chosen" are certain that they are in league with God.

Cohesion and the Psychology of Affiliation

Individuals who become affiliated with particular groups that support their own value system achieve a sense of greater security and peace of mind. The person's need for affiliation is one that seeks reciprocal validation of identity (you and I are alike), and gathers such validation partly through the ego reinforcement that membership in a group offers. Thus, according to Shaffer and Galinsky (1974) and, Tobach and Schneirla (1968), identity-affirmation becomes increasingly secured through the bonding of members within the ever-developing group cohesive social structure. And within intro and extra-punitive groups exists a muted but accepted euphoria because of the certainty that God or the God-head approves.

Galvanizing of Group Emotions

Yalom (1970), proposes that the group's cohesion offers all members a common set of curative factors such as support, hope, interpersonal feedback, and a sense of universality. In order to maintain specific group attitudes, affiliation in the group requires that a set of specified expectations of affiliation have been met. Usually tacit screening of membership creates the conditions whereby members are more likely to be obedient and accepting of group goals. In fact, Bion (1959) proposes a theory along with his group emotion theory that he identifies as "assumptions" around which the group coheres. These so-called assumption groups also generate typical agreed upon emotions derived from the motives inherent in their particular assumptions. The point of Bion's work is that it shows how underlying assumptions of groups can drive the group's motivation, while at the same time asserting basic emotions of the group, consequently sustaining the group's agenda. Bion correlates the basic group culture of these assumptions to six basic emotions— guilt, depression, hope, sexuality, hate, and anger. These emotions correspond to the assumption groups identified as: "dependency group" where *guilt* and *depression* are central and where the membership looks to the leader for security; "pairing group" where *hope* and *sexuality* characterize the basic underlying theme of the group interaction, and where the membership pins its hopes on some pairing relationship of members; and, "fight–flight" group where hate and anger determine interactions of the group and thematic material of extreme hostility or phobic withdrawal is typical of the overall group interaction.

The example cited here of Bionian theory underscores the idea that social psychological and group process understanding is in general agreement in positing an underpinning or unconscious organization to any group agenda. Such understanding is entirely relevant to our concern here with the relative punitive template of group culture and its affect on the cohesion of groups. Furthermore, such underpinnings to group behavior (also relevant to animal groups where hierarchical patterns are seen), seem to improve the group's survival capacity. With regard to human group analysis a glimpse of biological evolution can be noticed whereby comparisons can be made between animal and human groups. Of course in human groups the culture of the group will be far more complex. For example, extrapunitive assumed group cultures can be (although not necessarily) most punitive, intropunitive assumed group cultures possibly less punitive (although sometimes not), and impunitive assumed group cultures will in all likelihood be relatively free of blame or any punitive impulse whatsoever. In the impunitive group especially, cohesion is better established when members shed their need for so-called persona cover-ups and become more circumspect more candid, and even more introspective. However the cohesion of such groups is almost never as intense as that of the extra- or intropunitive groups because in the impunitive group there is really no imperative at all to idealize the group (even though such idealization may occur usually in the form of love and respect). And to idealize the group, its goals and its culture is the stuff of which cohesion is made.

In the extrapunitive group, where group idealization is highly important, a member will contribute to the group's cohesive structure when superego formation (the important element in blame psychology) as well as affiliation needs are based upon power concerns ("I am better than you and I will prove it by defeating you."). Along with this, covert feelings of inferiority and corresponding compensatory urges frequently require a violent or aggressive solution toward others. This is so because individuals of such groups need to suppress and then seemingly overcome feelings of inferiority or vulnerability and they are inclined to do this by way of social conquest. With such individuals, introspective traditions are eschewed partly because of the person's underdeveloped ego that is likely to be collaterally brittle. In addition, with respect to such group affiliational features such as the presence of compensatory needs, are included prestige attainments (Henrich & Gil-White, 2001), and a corresponding increase in status and pride experienced by holding membership in one's chosen group.

In an intropunitive group, members can enhance the group's cohesive structure when both non-accusatory superego force as well as affiliation needs become based upon security concerns as well as nurturant needs for protection—also against perceived untoward forces of the environment ("I'll be taken care of here and protected from those others.").

Further research into punitive group structure may ultimately reveal that the extra- and intropunitive types are interchangeable depending only on the cultural conditions that call them forth—and this would be an example of an epigenetic phenomenon appearing and asserting itself in the workings of groups; that is to say, that in culturally punitive groups—with respect to their so-called genetic endowment—environmental signals can trigger violence toward others in an ever ready, although dormant collective impulse residing in the coalesced membership. In addition such groups offer members a profound and revitalized interest in life, and finally, a sense of belonging to something they consider of maximum value— essentially validating the importance of their lives.

Thus, in examining the nature and scope of group formation and function, it becomes possible to appreciate how the need for safety and security reflected in the organization of psyche and mind parallels the psychoevolutionary approach to God, and as a evolutionary example, biologically traced from the point of the transmuted homologous function of the animal's tail into the function of the advanced human brain. So too did psychological issues of the need for safety and security similarly migrate or traverse the same evolutionary path. Then, of course, the psychological possibilities in derivative behavior based upon one's thinking became quite complex—even to the point of the thinking apparatus having the ability to create spandrels as in the ability to harbor opposite inclinations in the same person.

The question becomes how the thinking apparatus derives a spandrel that makes it possible to harbor opposite inclinations each of which contains its own motoric motivation—its potential to action? For example, how can Robert Hanson or Dennis Rader be devout Christians, worshipful and reverential churchgoer, and at the same time evil miscreants. The answer lies in the operation of the psychological mechanism of "splitting"—a stunning example of a psychoevolutionary defense mechanism of mind and psyche.

The Psychological Mechanism of "Splitting" Makes It All Work

The psychological mechanism of "splitting" as applied to groups, for example, refers to the splitting off or separation of good features attributed to the personality of the group from bad ones that are then solely attributed to other groups. In the intropunitive group the experience of revitalization occurs because members can have the opportunity to exercise faith in the goodness of some external force, while seeing other groups lacking such vitalization, either because of no faith or of the wrong faith. Faith then becomes the anesthesia of anxiety. It is generated by identity concerns that seek tension reduction through expiation as well as through a sense of communion. Hanson and Rader, then can be both bad and good by utilizing the defense mechanism of compartmentalization in the achievement of the splitting phenomenon (Kellerman, 2009a).

This "splitting" or split-off phenomenon is also relevant to Back's (1951) position that cohesion can be reinforced by a number of factors such as affiliation needs composed of personal attraction to the group, the successful performance of tasks, and the level of prestige attributed to the group, and by implication, as suggested earlier, a cohesion index can be a reflection of the relation between group punitive structure and individual superego formation. Back's reference to the issue of prestige should not be minimized. Prestige is an experience, a source of pleasure that can generate inner peace, a sense of assurance, and an assumed sense of strength—even joy. And generally, these positive attributes become characteristically identified with the subject group, while all negative qualities are "split-off" and attributed to other groups—groups that are identifiable and potentially targeted.

In this way, the uniformity of individual superego formation among members of the group is probably the most powerful element in homogenizing the group's compositional alignment. Such collective uniformity serves then to validate the particular punitive cultural structure of the group as "good." This sort of adjustment of one group member to another (also in terms of the issue of prestige of the group as well as to the issue of "splitting") is also considered by Cartwright and Zander (1968).

The importance of affiliational needs of members of a group relates directly to the potential index of cohesion, especially when there is a conscious effort to reinforce and fortify the issue of homogeneity of membership needs toward the common purpose of affiliation so that, again, good qualities are relegated to one's identifiable group while the bad qualities are "split-off" and directed toward other groups. A plethora of such studies on the nature of the group culture and its underlying purpose was, in the mid-twentieth century noted by many researchers (Cartwright, 1968; Festinger, 1950; Glover, 1953; Lott & Lott, 1965; Singer & Fesbach, 1959). A very important finding regarding such affiliation needs and the culture of the group was identified by Scott (1965), who investigated the dimension of attraction in fraternities and sororities and found that group attraction was not even based upon any significant degree of interpersonal liking. Rather, Scott ascertained that liking for the group-as-a whole was more important than liking for individual members. The important implication here seems to be that in the search for affiliation, potential group members will join for some reasons that seem obvious and for others that remain covert, or even unconscious. And it may be this essential punitive structure in the very nature of the group—this "splitting" mechanism that acts as a tacit selection criterion for membership in the group and is at the core of the group structure—the thing that attracts the member to the group (we are the right ones, the others are the wrong ones).

Thus, in terms of a "splitting" phenomenon, members may coexist whether they like each other or not provided that individual superego formations of all correspond to the specific group punitive structure. Then it becomes one for all and all for one. The punitive structure of the group offers each member an immediate sense of personal gratification, pride, and overall emotional gain. The danger here, however, is that members of highly cohesive extra- and intropunitive groups are able, because of such "splitting," to very possibly erase ethical, moral, and intellectual considerations when confronted with the likelihood of achieving even the most modest of group wishes. And this is the argument also for the ability of individuals such as Robert Hanson and Dennis Rader to be both entirely untrustworthy and at the same time be worshipful of God.

It is a case of the psychoevolutionary approach to God from a strictly "spandrelian" viewpoint whereby a socio/psychological perspective implies that on the one hand the individual can variably play the role of the worshipper and on the other hand also play the role of the omnipotent God who literally decides who lives and who dies. This spandrelian quality refers to a thinking brain in which options within the thinking along with the feelings accompanying such thinking can now generate vicissitudes of thought (and motivation) not originally intended as a component of such thinking—all of it constituting the cognitive spandrel in action.

With respect to affiliation, as far as the group is concerned, the group's goals serve to successfully offer each member an achieved sense of empowerment. It is empowerment with God's blessing or with the blessing of a Godhead. In addition, in the group, it seems that the subject, the person can be both a worshipper and a

God. This is so because of the apparent fusion of individual with group, along with the operation of the psychological mechanism devoted to compensatory behavior occurring within a "splitting" process. In such a case, the affiliation with God (as well as with the group) is one in which the search for such an affiliation is realized especially since in the "splitting" process, God is granted by man all that is powerful.

The impact of such a highly cohesive group on its members is profound, especially in persons of low self-esteem and therefore in need of reassurance, or for those in need of family acceptance, or of family, period. In such cases high group cohesion offers an amplified or compensatory sense of security while simultaneously extracting or perhaps even extorting an unusually commanding sense of loyalty. Festinger (1950) adds that the higher the cohesiveness the more power the group has to implicitly demand conformity to its goals. These consequences of cohesion, especially with respect to the effects of "splitting," might also be defined as group impalement rather than as simple group membership. In fact, the group to which any individual offers allegiance need not be located in any one place. Loyalty and adherence to an ideology can form a group based on a federation of membership in which each member can be an island of one, but that in total and metaphorically, it is a group in the form of an archipelago.

When considering these issues of group composition and group infrastructure, it is the group culture and its corresponding affiliated membership that becomes the God. This occurs in a homogenous fusion with the Godhead leader of the group.

Given the affiliational psychology discussed here it appears that groups can tacitly issue license to act-out group values in the absence of restraint that would usually be based upon individual morality and ethics, and this is true of religious groups as well as any other group embedded in some ideological foundation. Thus, it becomes evident that the individual in any group is confronted by a formidable group-as-a-whole force and that increasing cohesion can therefore generate an impulse, a tendency toward extrapunitive as well as intropunitive structures in essence, governed by a "blame" psychology—a punitive urge.

In the face of such group influence and persuasive force, any formed group will need to create a circuit for itself with mechanisms that become etched in the fabric of the group culture in order to attenuate the potentially extra- and intropunitive propensities that may always be dormant but which can be activated as a force to be reckoned with. Of course the entire issue of whether the group will remain embedded in a powerful punitive structure depends to a great extent on the nature and persona of the leader—the Godhead equivalent.

The Effect of Leadership on the Behavior of Individuals

A group with a high degree of cohesion will also possess a leader whose presence is continuously and strongly felt by the membership. This sort of leader frequently acquires mythic proportions thereby providing the group with a so-called projected ideal. It is what Armstrong calls "mythos," but in this case the mythos is joined in its idolatry with "logos"; that is, such a leader is adored with regard to deep emotional ties (the mythos part) and also for defined characteristics such as eloquence, or pronounced and assumed rectitude, or even because of a belligerent disregard of respect for targeted others (the logos part). It is frequently seen that such a role is actually ascribed or attributed to God.

As stated earlier, one of the dangers of highly cohesive groups is that a charismatic leader can create the circumstances of fixation on other groups as targets. This sort of danger regarding a powerful leader is of course palpable in both extra- and intropunitive groups. Such a leader produces inspired and potent followership. An overpowering leadership can create an acute atmosphere of "homogeneity–hunger" within the membership (as in cults) in which individual freedom is traded for unanimity, and in addition, obedience to the leader can be demanded along with the leader's insistence on compliance with certain rigid rules. Thus freedom of emotional expression may be ostensibly permitted, but this expression is never different from that of the overall group agreement as to what is desirable or even actually permissible. In such a case, what any individual member might otherwise feel that is skewed from the group's cultural ethos becomes irrelevant and certainly suppressed.

The major problem with whether individual expression of feeling really constitutes true freedom of expression is thus raised with respect to highly cohesive extra- and intropunitive groups because such individual expression generally represents the Bionian unconscious joining of fantasies and emotions of members of the group. Individual expression in extra- and intropunitive groups, therefore, is usually the expression of group emotions, and such group emotions are approved of and given their reason for being by the leader—the Godhead, or, the presumed God. The truth is that in such groups, the leader is indistinguishable from the God—is the God. Individual behavior under such circumstance becomes part of the group performance (and as an expression of the personality of the group) and thus, also represents a usurpation of individual prerogatives especially since these prerogatives were willingly yielded on the basis of initial affiliation. This usurpation can also be referred to as contagion, suggestibility, conformity, compliance, supplication, or even unconscious, automatic submission. It is a usurpation of individual autonomy transmuted into what could be termed the "group personality" or even the "group mind."

The Group Mind

In the group personality comprised also of the sense of ideological unanimity, this image of the group (the image of its personality) also makes for the existence of a foundation for myth to develop within the group culture reflecting the "goodness" of the affiliated group, and by default, the "badness" of other groups. Kellerman (1979) points out that the group, in its accumulated history, even wants to create and then continuously invoke both myths and legends of the affiliated group, all of which serve to elevate and glorify the basic group mission—in the most general sense of an idealized and shared view of the group by its tangible membership as well as by others and as well, motivated by the intangible sense of admiration and subsequent desire to belong. And usually, in such groups, of course the leader is highly visible.

At this point, it becomes impossible not to notice that in the discovery of God with respect to a psychoevolutionary perspective, reaching toward God in order to satisfy the need for safety as well as the need for communion and identification, the leader of the group (especially a charismatic leader) can easily become such a transference figure. Thus, in extra- and intropunitive groups, the idealization of the leader and the group also act to nullify anxiety of members and in its place to galvanize members of the group to consider themselves not only to be "firsts among equals" within the group, but also in an overarching sense, to be "first" with respect to all others outside of the group, as well as to all other groups. Ultimately this can mean that should the aim of the group become one of hostility toward certain other groups, then members will be much more likely to possibly become engaged in hostile behavior that under non-group conditions probably would or could not exist. And in this sense Kellerman (1981) has concluded that the seeds of fascism and its corresponding autocratic and tyrannical impulse can, because of an underlying high cohesion group consensus, be activated as a group mandate for individuals to do things that in ordinary non-group conditions would not and perhaps could not be done. In the affiliation with God (with a Godhead leader), the increased probability exists that anything can be done.

The pathological implication here was implied by Arlow (1961) who stated that the heroic myth of such a group serves as an instrument of further socialization. This type of socialization becomes engraved in membership functioning as well as in the group personality so that a pathology of blaming, understood to be condoned by the leader (of secular groups), or by some authority stating that God (in religious groups), begins to consistently characterize the personality of the group. And parenthetically, this phenomenon of following the leader to attack a blamed target is also vividly seen in animal groups as in the example of a chimpanzee group leader spearheading an attack on another chimpanzee group (Jha, 2010).

In extra- and intropunitive groups the cohesive culture then acts to form the group personality. Individual collusion in an unconscious joining of membership constitutes the so-called "group-mind" and this "mind" then becomes the dominant ethos-producing reflection of the ideology of the group. The collusion is the complete circuit between individual superego aims of members with the nature of the group's extra- or intropunitive structure—with the leader as the personified Godhead. This kind of group mind function is analogous to the homologous transmuted function of the tail as a security and safety function that is also seen in group behavior of animals and ultimately into a further morphed thinking brain of Homo sapiens. It is a specific journey of bio/psychological evolutionary function.

Groups need reasons to exist and for extra- and intropunitive groups such reasons are often characterized by an imperious focus on other groups. It is the reason that highly cohesive groups, as discussed by Kellerman (1981), can become, in their intergroup attitudes, hungry for targets. It is often that such a God-ascertained blame psychology becomes emblematic of the group.

In a general sense, a blame psychology is typically revealed as an inherent feature of the human personality; that is, especially when individuals begin to homogenize personal ego with that of the so-called group mind, surrendering personal autonomy for the sake of affiliation with the group ethos—its ideology and overall belief system. The point at which such an affiliative possibility is recognized is the point at which the line is crossed and acquiescence to a reinforced belief system (fortified by a group identity) is accomplished or rejected in favor of the autonomy of personal ego. These contrasting choices can also be seen as between homage to a presumed greater force (God), or some other ideal, and validation of an unshakeable belief in the inexistence of God.

As mentioned earlier, in considering the psychoevolutionary approach to God, it has been necessary in this chapter to examine the emotional/psychological component of the evolutionary journey from the protective/guarded function of the tail of animals to group functions and ultimately to the complex function of the thinking brain. An examination of the relation of individuals to groups could shed light on the psychological mechanisms (both abstract and concrete) utilized in order to achieve emotional security and sense of balance especially with the group as the significant variable. Parenthetically, in such groups, the acquisition of security through membership will, of course, supplant consideration of fairness toward others.

E. O. Wilson, in his volume *The Social Conquest of Earth* (2012) also examines the key concept of group affiliation as an evolutionary product. His thesis promotes the sense that evolutionary development reveals that groups seek dominance over other groups, or superiority over other groups, and will devise strategies as well as attitudes to support the instinctual drive toward such aims. Blame psychology can be such a strategy.

In the following chapter a further psychological explanation that may provide support for either side of the argument—the existence or inexistence of God—will essentially constitute a corresponding discussion of God and the core element of the human personality—its psyche.

Chapter 5
God and Human Personality

The human personality relationship with a belief in God is a bifurcated one. On the one hand we find most people who rely on their faith in God as a truth and not as assumption, and on the other hand there are those who rely only on what they consider tangible reality without any belief in a supernatural existence of God whatsoever. Of course there are many in-between shades of faith and/or belief in God, but for all intents and purposes we will consider these two.

Both sides of the personality/belief relationship are represented in the psychological literature where the debate rages regarding psychological mechanisms that can explain faith without relying on the concept of God (and for all intents and purposes actually supports the refutation of a God), and those arguments that claim precisely the opposite—that psychology and the study of personality mechanisms of the psyche, do not refute in any way the presence of an actual extant God.

The God Illusion in Psychoanalysis

In the psychoanalytic realm chiefly represented by the work of Sigmund Freud, God is understood to be illusory and a projection of the person's wish for safety and security. In this respect, Meissner (2009) stresses the difference between one's psychic reality and the validation of an actual and tangible object; on the one hand, imagining God expresses the subject's psychic reality, while on the other hand, validating a real divine being posits something about the real world.

Freud describes personality features (neurotic features) that contribute to the belief in a supernatural God. Freud (1927/1961) suggests that religious belief systems are essentially equivalent to obsessional neuroses. Further, such belief invokes defense mechanisms for the purpose of neutralizing tension arising from the uncertainties of life and a related fear of death. Obsessional neurosis as it relates to religious belief is a diagnostic conception that refers to the person's need to control personal anxiety by binding such tension with rigid limits.

H. Kellerman, *The Discovery of God: A Psychoevolutionary Perspective*,
SpringerBriefs in Psychology, DOI 10.1007/978-1-4614-4364-3_5, © The Author 2013

In this Freudian understanding, the obsessional part of the underpinning of belief concerns an attempt by the individual to control anxiety by searching for, and constructing order and precision—all of it for the purpose of gaining control over the environment and therefore to assuage personal tension. Freud's focus on the issue of "the uncertainties of life," and the corresponding "expectation of death," is also an oblique reference to the so-called death instinct that presumably drives individuals to think and do certain things toward the obliteration of anxiety. In the notion of "the end," in death, ostensibly there is no anxiety.

Freud's position is that belief in supernatural phenomena (such as a belief in God) is fueled and sustained by obsessional manifestations in thinking, feeling, and their counterpart—compulsive behavior. A repeated dimension here is how the psyche manages anger. In this discussion (Kellerman, 2008), it is proposed that the purpose of the appearance of obsessional states (or for that matter other symptoms as well) is to manage underlying dissatisfaction and anger. Underlying anger specifically, and dissatisfaction generally, exists in the person's unconscious because of the aforementioned uncertainties of life. And in many cases the issue of uncertainty actually implies the more fundamental issue of unfairness and the disappointment people experience about it.

The main issue here is that when people are troubled by uncertainty and vulnerability, and when they consequently feel that their wishes will be thwarted by unexpected variables in life, they correspondingly feel disempowered (or at least threatened with the condition and subsequent sense of disempowerment). The further psychological principle regarding such disempowerment is that anger is always, and without fail, the individual's consistent reflex to disempowerment. The reason for this anger-reflex to feelings of disempowerment (and consequent anxiety about it) is that when someone feels such disempowerment, then being angry is frequently the only way to feel reempowered. Therefore, there is a psychological reason for the psyche to automatically create compensatory defenses such as obsessional states especially when (for various reasons) the anger or dissatisfaction must remain repressed and out of consciousness. Thus, the reason the psyche generates obsessional states is to manage the anger that resides in the unconscious (Festinger, 1950). The anger represents the person's anticipatory existential sense of disempowerment covering everything from simple disappointments all the way to feeling impending doom—all of which is underpinned by the larger realization that life is unfair.

The logical extrapolation of such feelings of consistent disappointment and/or feelings of doom reflects the concrete institutionalization in the personality of how disempowerment and accompanying anger is managed in the person's psyche. Prevailing psychoanalytic wisdom is that such disappointment of wishes (thwarted wishes) and consequent anger becomes safely ensconced in the unconscious by the operation of repressive forces of the psyche. These feelings of disempowerment and anger that linger in the unconscious, inexorably radiate out from "below" and are then translated into obsessions or anxiety that people feel in the "above"—what they feel consciously. Beyond the symbolic meaning of the obsessional state, its objective is also to keep the anger repressed largely because anger or anger-mixed personality traits are considered to be unacceptable in civil discourse (Kellerman, 1979).

On a less symbolic and more tangible transferential level, Freudian philosophical metapsychology also posits that the image of God-projections are really of parent images that exist in order to ensure protection and a sense of safety (even salvation) that will then help the concerned individual perhaps transcend death, or at the very least, perhaps dilute the tensions of life—and to be successful in this respect even vaguely implies that the issue of "fairness" is at work.

Rizzuto (1998), in a treatise entitled *"Why Did Freud Reject God?"*, presents a similar interpretation regarding projection of the God image and cites Freud as one who insisted that God was nothing but the person's wishful emotional dependence on a revered historical childhood father that then gets transformed into another being—a supernatural being. Thus, Freud's only reverence was to "logos," or as he put it: "our God Logos—reason" (Freud, 1927/1961).

Actually, Aron (2004) refers to a Rabbi Soloveitchik, in a psychodynamic definition of God that is even more psychoanalytic than Freud's. In Soloveitchik's view, God arises simply as a projection of the individual's need for protection. Soloveitchik postulated that man is affirmed in his adoration of God and then because of it sees himself as a great being. Soloveitchik also proposes that man sees himself as a "nothing," and even as ruthless and inconsiderate—actually a negation of self. Thus, man is exalted and therefore grandiose in finding this great being (the self), while at the same time being profoundly submissive, as a "nothing" seeking protection, and may be not even worthy of it.

This then, is man's existential experience of his own esteem and is a focus even in a specific kind of therapy known as logotherapy which is based upon a search for meaning related to spiritual or social issues, aiming to achieve what is known as "will to meaning" (Frankl, 1969; Kellerman, 2009a).

Since projection as a defense mechanism is a direct outgrowth of a person's need system and wishes, then according to Freud, the concept of God is entirely a product similarly of the individual's wish fulfillment and need. Freud goes further to say that since people disregard the contrast between illusion and reality, it turns out that such disregard of reality elevates the illusion to the state of delusion. In other words in illusion one simply misinterprets something that is there (in Freudian terms the projection of God), while in delusion there is a false belief in the face of evidence to the contrary (Kellerman, 2009a).

Ultimately an intense illusion can become a denial of the person's intelligence, actually an intimidation of the person's intellectual ability. A false belief belies any sort of evidence to the contrary. Ernest Jones (1974) goes even further by relating the belief in God to narcissistic elements of the person's need system so that excessive narcissism becomes a motive regarding the conception of God. In other words, the stronger the need for self-validation and ego elevation in terms of the excessive need for examples of adoration, the stronger will be the need for projecting a God—for finding or discovering a God.

God and the Self

Freud (1901/1914) postulated that in a psychic dance of projective images, man says that God created man in His image but then the reverse occurred insofar as man created God in his image. Meissner (2009), in his book, *The God Question in Psychoanalysis*, also indicates that the God representation is modeled on one's severe superego. And the implicit reason is that people probably need a God who is strong enough to demand high-level "rightness" and that in the absence of such rightness, penalties can ensue. Of course, the severity of such a hierarchical force needs to emanate from a powerful agency, and that agency is the all powerful God; and even more so, potentially an all powerful and punishing God—the sort of God that likely invokes a punitive blame psychology in its interpersonal God/person relationship, or even God/group relationship.

Such a God/person relationship is also a reflection of how the individual sees the self. The need for restraint and self governance will usually need an outside force to police one's impulses and this force is what Freud considered was projected onto another object force outside of the boundaries of the self. From a psychoanalytic Freudian perspective then, and as a result strictly of the defense mechanism of projection, this object force is deduced and conceived as God. In mathematics, this is akin to creating a variable that is outside of the "box" in order to solve the problem within the "box," discussed in Chap. 2.

The problem within the human parameter (within the box) concerns tension about one's impulses, needs, and hungers (of all sorts). Thus, in order to successfully navigate demands of normal socialization, the individual "knows" that any assistance would be welcome, or in fact is needed in order to tame these impulses as well as to tranquilize one's personal insecurities and feelings of helplessness. It then becomes necessary to construct ways of containing and then also restraining such less-than-civilized impulses that are always struggling to be released, and correspondingly to create mechanisms that can manage perceived untoward possible events in the world.

Hence, what is implied according to the Freudian sense of man's internal and perhaps chaotic experiences is what needs to be projected onto this proposed external object force. Perhaps the image that is projected (and remains outside of the "box") is one's externalization/projection of God. With God as the ostensible or presumed extra variable to assist in the restraint of impulse, Rizzuto (1996) considers that such a God representation affects the sense of self. This is true because of the believer/God relationship that in turn influences how the individual perceives the self in view of this faithful relationship of self with God. It is the same theorized self-imposed method of giving oneself empathy by feeling understood by God. In this case it is elevating one's stature by this hypothetically self-constructed method—a belief in God's presence.

In another reference to the relationship of God and the self, Winnicott (1965) suggests that what is good is attributed to God in order for the "good" to be free of contamination which would not be the case if the good was only attributed to the self. And this is so because the person understands intuitively that at the most, quite consciously, that the life-of-the-self is rather like a condition with a host of ambivalences, contradictions, imbalances, insecurities, illegalities, behavioral anomalies, and anything that can be defined as those emotionally unrelenting experiences that challenge one's moral posture. This challenge of one's moral posture very often forces the person to choose convenient short-cuts that, in essence, are ethically rather thin, or actually ethically questionable. In addition, because of such possible contamination of even the good things, questionable judgments made can be of the second order rather than good judgments chosen that would be of the first order.

Therefore, because of the person's sense of self as an ostensibly compromised creature, all "goodness" not only seems to be better protected when ascribed to God but is felt to be definitely completely protected when ascribed to God. This same idea is unfolded by Feuerbach (1941/1957), in the book *The Essence of Christianity*.

Meissner sums it up by saying that whether it is the protective father or what he calls the security-enhancing selfobject, God is still one who emerges from one's needs and motives. Yet, Rizzuto also considers the possibility that the illusion of God conceived by Freud also can be rejected in favor of asserting that the individual, actually, in addition to perhaps alluding to an illusion of God nevertheless perceives the external reality outside of the psyche to contain evidence (spiritual evidence) of the actual presence of God.

Can a Presumed Extant God be Refuted?

The general argument about whether it is God that is being discussed or simply a God representation (as in a projection of the person's need for security and safety), in either case the experience of God can also be discussed in psychological terms with respect to the psyche's defense mechanism of compartmentalization as was exposited in Chap. 2 and Chap. 4; that is, that God becomes the final arbiter of right and wrong, and good and bad, simply because the person assigns the inferior status to the self and the superior one to God. This sort of "splitting" permits such compartmentalization to occur. Nevertheless, even in view of such psychological explanations of the presence of God, or of a God representation based on needs of the self, none of it ever can prove or dispute the possibility of the presence of an actual extant God.

Even with Freud's definition of criteria that characterize the unconscious so that should the projection of God be shown to arise from self-needs for protection, these criteria still cannot be an absolute refutation of the possible and actual existence of God. For example, Freud discusses five characteristics that together reveal the unconscious domain. These are *timelessness* (time in the unconscious is not an evenly spaced meter, nor is it only experienced in forward motion), *spacelessness* (wherever one is located at any given time can be quite ambiguous), *displacement/ condensation* (one thing can stand for another and subject matter can be compressed), *equivalence of inner and outer reality* (for all intense and purposes there is no unconscious difference between the two), and *lack of contradiction* (any two seemingly opposite conditions can exist simultaneously without the psyche being at all concerned). Thus, even in the unconscious, no matter how the issue of God is treated with respect to time, space, symbolism and compression of feelings, as well as the presence in the unconscious of mentation that is illogical and even surreal, and as well as a lack of concern in the unconscious regarding contradiction of the existence of any number of elements—none of it, either separately as in component by component, or as all combined, can at all refute the notion that a God might actually exist, and this is true even though the theoretical network may be especially proven valid with respect to the entire personality system, as well as because of its explanatory power that also even may be considered—irrefutable! Theoretically irrefutable or not, theists as well as secular atheistic scientists will agree that none of it can actually disprove the existence of a God.

And in the psychology of the self, even theories that posit the projection of God as an attempt to control the individual's grandiosity by attributing such grandiosity to an external force—as a way to distract oneself from seeing that the grandiosity and wish for omniscience is really a product of the person's psyche—nevertheless, though such grandiose wishes of the self might be inherent or even perhaps genetic, all of it still cannot absolutely refute the possible existence of an extant God. And even with Grotstein (2000) introducing the idea of projective identification as the direct psychic mechanism that makes possible the psychological relation of self to God, nevertheless such an idea which posits that the person projects the wish for a God actually to be there—in extant form—and then the person identifies with this tangible God, this also cannot be a definitive refutation of the presence of some external "real" God. Meissner (2009) joins the argument regarding the relationship of the psychology of the self and belief in God by stating that belief is connected with a coherent idea of one's identity.

What such theorizing can do is to present a powerful matrix for understanding (in the Freudian case) that God is simply a projection of issues and elements in at least the psychological constitution of the self regarding concerns of survival that always need to be assuaged. Yet, absolute refutation of God remains impossible, even with the power of such encompassing theories of the mechanisms of the human psyche and its derivative adored "psychological self." In fact, the opposite argument from that positing either an illusory God or one that is object representational and therefore projected from the psyche is that enunciated by Spero (1992), who proposes that since God exists outside of the human mind, that to think of Him as only illusory or representational of the psyche is, by definition a false or even illogical conclusion.

God as Transitional Object

An interesting issue in the discussion of God and the human personality concerns what Winnicott (1953) called the developmental phenomenon of the "transitional object." That is to say that in child development this transitional object as defined by

Winnicott can be generally characterized as an external object that the infant adopts or creates which then acts as interstitial tissue (a bridge) between the infant's complete dependence on the caregiver (mother), to only a relative dependence on the caregiver, and on to separation from the caregiver into a greater sense of independence and autonomy of the self. Within the process of such development the transitional object becomes a rather soothing agency.

The theoretical position of God as a projected wish-fulfillment phenomenon and as a mirror image of the individual's ideal self, permits identification with the person's own projection and is what enables this rather soothing experience to occur. Such an understanding suggests that each person in their own projection creates their own particular profile of God as transitional—in the sense of this example of the transitional object easing the pressures of early child development. Rizzuto (1979) amplifies such a proposition by referring to the "ego-syntonic God." In psychological terminology, "ego-syntonic" is a reference to anything that is acceptable to the ego so that in relation to such an ego-syntonic object, anxiety becomes attenuated and the subject can feel safe (Kellerman, 2009a). Rizzuto postulates that the ego-syntonic God is roughly equivalent to mother/child interactions and experiences, and such a concept qualifies as supporting Freud's notion of God worship as an incubation and development of the psyche's need for illusion in order also to enable the individual to feel safe and free from anxiety.

Winnicott also argues that the transitional phenomenon creates the capacity for illusion which gives man the corresponding capacity even for the creation of religion and religious experience—a thought which inherently contrasts with that of Wade (2009a, 2009b) who considers identification of God as wired into genetic DNA. In any event, the capacity for illusion is also an attempt to create total fairness in the sense of gaining security, safety, and balance in this unfair world.

The furthest extrapolation of the issue of God as a transitional object could be considered postulated by De Mello Franco (1998), who in discussing religious experience and psychoanalysis concludes that a nonreligious condition is not in the least a matter that should interest psychoanalysts because the nonreligious condition already represents a kind of liberation from the childhood illusions of the infantile mind. In other words, freedom of the presumed constraint of religion is essential proof that such religion (and also the belief in God) has been, so-to-speak, outgrown, since by definition a transitional object is no longer of any use in the person's fully autonomous adult life, and where therefore the ultimate concern with fairness recedes in favor of the facing of reality.

Since individuals determine what their transitional object(s) will be, therefore, when (and if) such a transitional object becomes the belief in a God, then such belief will always be idiosyncratic to that particular person, and so the definition of God—how God is seen by different people—will reflect the difference in "the variety of needs" of each person; that is, assuming that there actually exists a "variety of needs" rather than a homogeneous cluster of needs germane to all people, and for that matter, possibly for all animals throughout the phylogenetic scale. The projection of such needs, in a process of projective identification (project it and then identify with it) and object representation (the mental representation of an object) are then correspondingly seen as familiar by the subject or simply, identifiable. Basically, as Meissner points out, what this means is that each believing person creates their own image of God gestated from personal needs.

The Zeal Toward God and the Context of the Supernatural

In the sense that each person may create their own image of God based on the presence of psychological mechanisms, such subjective images of God also usually gain communal support from like-minded individuals that negate a synthetic, artificial or man-made reference to God and rather reinforce and fortify the Godly certainty of God. This is true especially with respect to the sanctity and subjectively feeling rightness in the certainty of God's existence. To this certainty, Bloom (2005) explains that religion brings people together and can be palpably experienced as a social glue. It is the sense of fraternity that also explains the possible extrapunitive nature of such like-minded people in their particular religious subculture that, in addition to doing good things, can target as "bad" those who do not share the core faith of the group, and as Bloom suggests, also where the group may reserve, and even assert their ire toward those considered apostates.

Along with membership in such an extrapunitive group is a zealous God-fearing subculture where rituals and sacrifice deepen the loyalty (and even stridency) of the group. It is a group in which the entire membership helps to consolidate deeper group cohesion. As suggested earlier in the discussion of cohesive groups (Chap. 4), when such deep and intense cohesion exists it gradually begins to metamorphose into adhesion rather than cohesion so that extra-natural explanations become possible especially if such ideas further the aims of the group's ideology—in this case a zealous belief in God.

Examples of such "unusual" extra-natural ideas include perceiving the mind as being capable to envision all sorts of unexpected predictions: to wit, as the example Bloom observes in the idea of soul-less bodies and bodiless souls. The ability to see the mind this way—that is, in the mind's ability to imagine unverifiable phenomena such as creating logic in order to believe in possible but unverifiable phenomena (an afterlife, or even God)—almost leads inexorably to the pronouncement of the theory of creationism.

As much as God-fearing worshippers will debunk the psychoanalytic Freudian notion of a projection of God (that nullifies the notion of a real God), so too do atheists debunk the idea of what they consider to be the extra-natural idea of creationism.

In the study reported by Cathy Lynn Grossman (2007), and carried out by the Baylor Religion Survey (referred to in Chap. 3), the presence of a guardian angel, and hearing the voice of God were noted as palpable experiences by significant numbers of people. In addition many felt they were being looked after by a deceased loved one. In addition, Kenneth Woodward in a Newsweek article (1992), entitled *Why America Prays*, picks up on this "unusual-idea" theme and along with a similar discussion by Michael Shermer (1997)—in his book, *Why People Believe Weird Things: Pseudoscience, Superstition, and Other Confusions of Our Time*—lists and describes supernatural beliefs that more or less provide support for a context in which God or a supernatural world is not only possible but is believed to absolutely exist. Some of the practices in this "unusual" belief system include seeking psychics to predict one's future; use of Iridology to diagnose pathology by examining the iris; astrological predictions of the past, present, and future; extra-sensory perception that may give one the ability to read minds; communication with the dead; a belief in poltergeists or ghosts; the Bermuda Triangle as a presumed catastrophic environment waiting to engulf people; levitation or the ostensible phenomenon of magically nullifying gravity; the use of psychic detectives; a belief in haunted houses; the practice of astrological birth control; and a belief in alien abductions.

To sum it all up, Robert Wright, in his book, *The Evolution of God* (Wright 2009a), provides a seemingly iron-clad logic to this element of popular culture expressed as a zealous attachment to a belief in the supernatural. Wright indicates that to explain religion (implying also popular insistence on belief of an overall supernatural context), we need to appreciate that the human mind is amenable to all sorts of beliefs and practices. The seemingly undeniable answer is that the mind (as at least one product of the biological brain) is always amenable to needing and wanting to be safe and sound. Rather than a genetic connection to the belief in such a supernatural context (including, but not limited to, a belief in God) is Richard Dawkins' (2006), idea of culture as a triumph over biology. Here, the zeal toward believing in the supernatural (or extra-natural), refers rather to cultural evolution; that is, the transmission of belief from one person to another (and from one generation to another) with respect to all kinds of belief systems as a way to soothe the fears of individuals as such fears relate to life and death, and to allaying the anxiety that people experience in day to day uncertainties and "unfairnesses."

According to Dawkins and others, the insistence on belief in a supernatural context (including in the reverence of God), offers people a hope and feeling of having greater control over their environment. As such, we can see that in more primitive cultures, that is to say cultures without a formal humanities and liberal arts educational context and without any significant industrialization, individuals will almost ubiquitously invoke the intervening God to explain why bad things happen, why good things happen, and also to ask for the bad thing not to happen and only for good thing to happen. This can also mean that people equate fairness only with good outcomes and unfairness only with bad ones.

Even in psychological diagnoses of individuals, those who are more identified by hysterical features as their personality centerpiece, will be more suggestible and more easily given to supernatural explanations. This may be due to feelings of ineptness and highly emotional states generating an atmosphere of hope.

No matter the road to the supernatural: survival; peace of mind; a sense of security with greater assurance of protection from above; and the need for emotional balance, what is sought in all of it, is the quest for certainty. With such a significant need for certainty, people can easily confuse the issue of causation with that of correlation. And further, individuals who are in acute need of certainty will usually actively seek to conflate the two so that it becomes more possible to validate the supernatural or extra-natural. For example, a person is thinking of a friend who lives across the country and at that precise moment the phone rings and lo and behold that same friend has called. The receiver of the call who was the one thinking of the friend then may feel for sure that there was some sort of telepathic connection—a synchronicity—correlating the "thinker" with the friend. In such a case, the "thinker" might not consider the correlation but rather reflexively consider the event as somehow and in someway causative. The interesting phenomenon here is in assuming it was a coincidence. It is clearly quite easy to be seduced by such an amazing event. However, to feel a compelling influence (other than coincidence) that transforms the entire experience into a sense that after all, it was no coincidence at all, but rather causative or synchronous is a seeming conflation of meaning or even of thinking. Of course, the receiver of the phone call may have thought of the friend numerous times in the past but never in those previous times had the friend ever called. But to those who need the support of the extra strength provided by ostensible extra-natural phenomena, it only counts when it works and therefore is then remembered. Those who are compelled by such an event may insist on its causative certainty, which from a scientific vantage point and with respect to that one particular event really cannot be decisively refuted.

The issue of connecting correlation with causation has been in the past practiced with horrible consequences. For example, Hitchens (1987) points out that the Aztecs would each day tear open a human chest cavity in order with such sacrifice, to encourage and entreat the sun to rise. In this way the authorities in such cultures (that in contemporary time would be considered primitive) were able to create the circumstances for people to believe (to be terrorized to believe?) that there was a contiguous relationship between the tearing open of the chest cavity and the assurance that therefore the sun would rise. Boyer (2001) suggests it is such thinking of the supernatural that exists and receives its reason for being, only and directly from the projection of the mind of man. In other words, presumably if there is no projection from the mind of man onto the external world, there would be no supernatural at all, including no idea of God—and to boot, no one's chest would be torn open—each day!

In summary, we may say that human personality has the potential to activate a variety of psychological mechanisms to permit a variety of thinking avenues to be constructed or construed in order to establish safety nets that offer individuals reassurance regarding any number of tensions and personal concerns, and therefore that human cognition is able to project onto the external world the idea of a God to whom one can turn for succor and generally, for a sense of safety.

And still in all, no matter the presence of such psychological propositions, in no way can this be proof positive that a real God is inexistent.

Chapter 6
God and Belief

The More Subtle Definitions of God

In the definitions that have been applied to God by various theologians, scientists, and an assortment of other interested parties, such definitions range from the commonsense to the intellectually driven phenomenological ones, all the way to the highly esoteric. In most cases, the absolutists are those who offer the most commonplace definitions acknowledging God's existence—take it or leave it. On the other hand, intellectually minded philosophers and theologians who are life-long believers or who have spent significant time thinking about God, offer definitions that seem to have been formulated in especially original ways. Such thinking emerges as intelligently egalitarian, ever interesting in their Godly definitions, and hypothetically could not immediately be dismissed as wrong. As such, original thinking by Freud might be considered by like-minded nonbelievers to be absolutely correct and proof positive that a true God does not exist. So too, original and sophisticated thinking by theists might be considered by like-minded people also to be absolutely correct and proof positive that a true God does indeed exist.

Woodward (1992) reports some of these newer ways of defining or thinking of God. One definition regarding how one might think of God includes the idea that one cannot think of God at all because doing so does not correspond to Godliness. This is quite similar to Karen Armstrong's reportage of the history of the search for God (2009), insofar as she reports that historically there are those who say that *God is nothing* because God is not another human being. Others imply therefore that it is only *not* thinking of God that defines the relation between God and events concerning Godliness. Armstrong, assuming God as real, portrays the search for God and posits that the encounter with such a phenomenon is defined as the experience of a sacred reality. From a psychological point of view, the belief of God "as real" perhaps in a fundamental way gains its energy from the need to symbolize and even embody the image of the good, loving, and ever present protective parent.

The Jesuit Father, Dick Rice, founder of *Loyola*, a spiritual church center in St. Paul says that talking to God should be deferred in favor of listening to God.

H. Kellerman, *The Discovery of God: A Psychoevolutionary Perspective*,
SpringerBriefs in Psychology, DOI 10.1007/978-1-4614-4364-3_6, © The Author 2013

Then there is the Trappist Father Thomas Keating who founded *Contemplative Outreach*, who wants people to repeat some sacred word in order to center the mind, presumably in the sense of increasing the probability of accessing a reverential communion with God. And then there is Terry Eagleton (2000), who believes that God is anything and everything and more or less, the possibility of any entity whatsoever.

In addition, Herbert McCabe (1987) says it all as in the title of his book: *God Matters*, as does Dean Overman in the title of his book: *A Case for the Existence of God*. Furthermore, Mark Johnston (2009) in his book, *Saving God: Religion After Idolatry*, indicates that God cannot be made as an ally and can only appear as "…. the wholly Other, the numinous one [divine] who transcends anything that we can master by way of our own efforts." With respect to the type of God people worship, Johnston's God is not necessarily an intervening one, not necessarily impartial (or nonintervening), not necessarily irrelevant, and of course not at all inexistent. In fact, Johnston states that most religious belief is idolatrous so that God is "the Highest One which is equal to the Outpouring of Existence itself." In addition, in this conception, God is the Outpouring of Existence by way of exemplification for the sake of the self-disclosure of Existence itself. Thus, and in other words, without God there is no existence at all.

Further, the philosophy of Alfred North Whitehead (1861–1947) is known as "Process Philosophy" or "Process Theology." Here God is not omnipotent but existing by free will, as the universe changes in its ever moving process. God, actually, according to the process, cocreates with people and therefore God can only influence and not dictate. And in this sense, God is only a bit of an intervener but only perhaps with gentle persuasion.

And then there is the philosophy of Michel Henry exposited in the book, *I am the Truth: Toward a Philosophy of Christianity* (2003), in which Henry considers that God is life itself. According to Henry, divine life is real, and so God is not merely the divine Being—he is logos (the Word itself), who is the formal element of this Being. Of course as the "he," perhaps Henry makes Him a him.

In Fleming Rutledge's book, *Help My Unbelief*, (2000), God, according to Rutledge is "….who he is in himself." And in citing Sarna (1986), Rutledge references Sarna's reflection and understanding of God as "….the transcendent, awesome, and unapproachable Divine Presence" and in Rutledge's words, (God is) "….self-generated, self-perpetuating, absolute and underived, unaffected by anything outside itself". Perhaps the operative and profound term here proposed by Rutledge is "underived." In this same spirit, finally, Paul Tillich (1963), quoted by Armstrong (2009), states "God does not exist. He is being itself beyond essence and existence."

In Jewish tradition, access to God is relational insofar as there is a mutual and intersubjective relationship between God and humanity (Aron, 1996). And a covenantal relationship requires mutuality not symmetry. Man is not symmetrical with or equal to God. This is so because the comparison between man and God is essentially hierarchical. However, because of such mutuality, man is then able to question

God, and even to put God on trial (a Din Torah or trial of God). According to the Talmud, being good on earth will be rewarded and so the implication leads to the proposition that it may not even be necessary to praise God. Thus, in Judaism, members are taught to question and confront God—in essence to maintain a dialogue with God—holding God to the fairness principle.

Aron (1996) indicates that the God of Maimonides, following Aristotle, is abstract, even beyond the need of anyone—perhaps an oblique reference to God as not merely impartial, and also as far as the Earth is concerned, possibly also, in a way, and in the absence of malice, irrelevant.

Then there are other so-called esoteric definitions of God. Because of the unusual generality of such definitions (perhaps their subtle unsubtlety), they create the most overarching consideration of the presence of such a possible deity. Such general definitions can render the meaning of God, for all intents and purposes, to lose its specific configuration, as for example in the Johnston definition that equates God with "the outpouring of existence," or even that as a noetic (intellectual) idea, God might exist but only in the arena of the human brain/mind that is not understood.

Finally, as an example of extremes, the Christologists of the late twentieth and early twenty-first centuries created a new sense of subtracting the image of Jesus as a God while focusing on Jesus the man. Mordecai Kaplan, the Reconstructionist Jewish theologian prefers to focus on the "idea" of God in contrast to seeing the reality of God or, in fact, even considers a possible absence of God. Therefore, Kaplan, simply accepts God as an idea because he feels that in a rather basic way thinking about God helps people. Thus, it may be that Kaplan would not be insulted by Freud's notion of God as projection or even at the idea of God as a mechanism of the God-power assigned to the externalized "box" that helps solve the concerns of the individual's internalized tensions. Of course at bedrock, Karen Armstrong simply believes that God is transcendentally indescribable.

God: Outside of the "Box"

Another way of understanding how man solves his tension regarding survival and the need for protection is to adhere to a faith in God fortified and validated by reliance on scripture as well as on the reinforcement of such belief by parental figures. In this sense, a person can feel that God is on the outside—a figure external to the self—looking in or, from above, looking down at human beings. Thus, God is seen as an object outside of, or external to the worshipper. This separate "being" can be considered a phenomenological mechanism construed, as it were, outside of the "box." What is in the box is the corporeal person and immediate reality in which such a person exists. The God-power outside of the "box" then solves the problem of people within the "box."

The problem within the "box" concerns universal tensions about survival, and the only way that many people come to the idea of solving this problem of moment to moment survival as well as resolving the dilemma of anxiety about life and death, is to discover the God outside of the self, and then to see that such a God is able to be in some form of communion with the self. This communion then acts to satisfy the yearning for protection and safety that is apparently a ubiquitous one for all living creatures, especially those creatures with consciousness and an advanced thinking brain.

Jerome Kagan (1996), the American psychologist quotes Johann Fichte, an eighteenth century philosopher who posited that what is good produces happiness and not the other way around; that is, that it is not happiness that produces goodness. This, of course, raises the age old question as to: Do the Gods love it because it is good, or is it good because the God's love it? If Fichte's adage is taken seriously, then it may be that God loves it because it is good, and so it seems that God learns from people/nature about what is good/natural. And this also seems relevant to the Christian teaching of God loving one before deeds, meaning that at the very root, God loves you because perhaps basically you are good—and ultimately made in the image of God.

Robert Wright (2009a) indicates that with social evolution man has become more moral and that religion has helped people know right from wrong. Modest intellectuals, who are theists, as well as most atheists, could take issue with such a proposition. The argument would be that empirically, with respect to large social phenomena, there is evidence to the contrary, as for example, in all the religious wars and continuing religious animosities that probably support the idea that with social evolution man has become less moral and that religion actually has not helped people know right from wrong. And further, with respect to large sociohistorical events, as for example in the centuries leading to the Spanish Inquisition in the late fifteenth century, and continuing even much later—as in the Christian focus that brutally targeted Jews as a result of a Christian highly cohesive extrapunitive culture—one could consider that religion actually contaminated the moral rectitude of individuals. Then the Nazi-led Holocaust occurring approximately 500 years later in the mid-twentieth century essentially targeted the same people—not to mention the early twentieth century Turkish genocide against their Armenian citizens, or genocides occurring in Africa and Asia, or the genocide against native Americans as well as the brutal centuries of slavery of Africans.

Thus, all of it, again, could indicate that quite the opposite is true—that with respect at least to large social events, and with social evolution, man has become not a whit more moral and that religion has not had the slightest effect in helping people to behave with respect to knowing right from wrong not with standing the "better angels" of Steven Pinks (2011).

On the other hand, avowed atheists would consider Wright's proposition about religion as an example of a bastion of egalitarian thinking, to be utter nonsense. But if we look at the effect of religion with respect to how its affected individuals, we can say that if we subtract the punitive and other morally repugnant attitudes of one religion against another and just observe the moral teaching and how religion may influence individuals, then we might be able to appreciate more fully Wrights opinion about the value of religion in society—at least to this growing edge point of evolution.

Philosophical Arbitration: Atheists and Theists

Atheists such as Richard Dawkins, Daniel Dennett, Sam Harris, Christopher Hitchens, Steven Pinker, Victor Stenger, and like-minded others, generally feel that belief in the supernatural (requiring the construction or projection of the external "box"), is not just false and illusory (as Freud proposed) but also dangerous. In aggregate they warn against religious fanaticism and cultism, such as extreme Christian fundamentalism, Islamic extremism, and extreme Messianic Judaism. And they also decry old or new ideas such as crystal-ball readings, palm reading, fortune telling, out-of-body experiences, alien abductions, levitational experiences that ostensibly cure illnesses, psychics, and the like.

Those that choose such avenues of rescue are also in a very profound sense seeking some sort of elevation in ego, or safety, or in the extreme, immortality. To this quest for rescue and subsequent continued existence, Silverman (2006/2007) responds that the seeking of immortality is for atheists replaced by the value of a person's good works that live long after the passing of corporeal life.

The atheist argument becomes the underpinning of the aphorism: "God loves it because it is good." In that sense it becomes necessary to examine the genesis of "goodness." Is it that knowing and feeling goodness fundamentally resides in the genes as a genetic given? Or is it that one's rearing and development during formative years in the sense of good early nurturing is the key to knowing good from bad? The answer most likely is an epigenetic one—that there is a genetic altruistic inclination based upon Darwinian survival imperatives that becomes confirmed by early nurturing within environmental conditions. This issue of environment and culture as essential properties of epigenetic phenomena was considered by Michael Meaney at the Neil Miller Lectures (reported by Michael Price, 2009), in which Meaney, reflecting Miller, makes clear that DNA itself is not the whole story—a point also presented by Confer et al. (2010), in their treatise of evolutionary psychology. Such conditions are really early experiences in socialization. If the socialization was loving, giving, and caring, and in addition made mature demands of the child, then such individuals develop the interpersonal as well as intrapersonal knowledge responding to an animated genetically based instinct enabling one almost automatically, instinctively, to know right from wrong.

Even more importantly, such epigenetic achievement enables one to perhaps calibrate decision-making processes based upon this consciousness of what is right and what is wrong. And this is presumably so whether the person is religious or God-fearing or not. In addition, with decent and normal developmental experiences, religion generally has the ability or assumed persuasive power to influence individuals in a positive way. By the same token individuals who have had poor early experiences may be influenced by this same religious exposure in a negative way, or perhaps not influenced at all. Psychologically, the prediction is that those who have been reared with loving and caring attitudes will be the ones most unlikely to be negatively persuaded by any undue imperious religious or other intro or extrapunitive inclinations.

Robert Wright's position is based on the hopeful assumption that love may be manifested from a Divinity (external to the person). In experiencing such love individuals become more like the Divinity—and to be more like the Divinity contains a moral modeling template, a moral compass. The underlying theistic message is that "God loves you before deeds." Thus, Wright's reference to love as manifested from a Divinity can be a persuasive positive force almost compelling the worshipper to be "like" the Divinity. The relationship between the worshipper-person and the worshipped-God is precisely what is meant by creating the external condition (the "box") that addresses internal dilemmas, and even answers internal needs and hopes.

Jon Meacham (2009), in a *Newsweek Magazine* essay states that: "….faith is an intrinsic impulse….The belief in an order or a reality beyond time and space is ancient and enduring." The argument that can be made here is that what is ancient, enduring and an intrinsic human impulse, may not solely invoke faith in God. Rather, the enduring, ancient, human impulse may be the wish for protection, security, emotional balance and safety, the expectation of fairness along with overall peace of mind, personal empowerment and the absence of any disempowerment, out of which is faith born.

The Meacham Newsweek essay quotes Charles Grandison Finney, a nineteenth century evangelist who says that the function of religion is to reform the world—"… to put away every kind of sin…" However, it is likely also that sin may be a function of a person's compensatory acts addressing what in that person's life fell short of protection, love, safety, security, emotional balance, caring, and overall peace of mind. In its deepest genetic/evolutionary meaning, it may be that to rid the world of sin is to ensure developmental safety and caring, early on in a person's life— from the beginning.

To continue this discussion of the conjectural thematic dichotomy—the theist/ atheist polarity—a question regarding faith itself could be stated: Without the need for survival and the search for safety, would there be any need for faith whatsoever? Would that mean that there would be no need for a God? The atheist might posit that in such a case no God would exist. The theist on the other hand might posit that God's existence does not at all depend on any man-made logic or logos, so that the presence of faith in itself or for that matter even survival needs of organisms, are not at all relevant criteria determining God's existence. Therefore, even though under certain safety assurances people might not have a need for a God, nevertheless, none of it can absolutely refute the presence of a God. Yet, in the absence of a Darwinian universal and ubiquitous need for security, safety, and survival, it would perhaps become less urgent for the thinking mind to posit the need for God. However, even this subtraction of the Darwinian need cannot decisively abrogate the existence of God!

The issue that may be hidden here concerns the possibility that whether God is real or not is truly not the issue. The real issue may be the quest for a loving, caring, and fair environment that generates in people a sense of safety, security emotional balance, an expectation of fairness, and overall peace of mind so that the possibility of a God's presence does not at all depend on the state of a individual's personality or fears. In any event on the face of it, whether it is a relevant God or not, it doesn't seem that such a so-called entity would have its existence based upon personality or fears, need for empowerment and nullification of disempowerment, or even genetic organization.

The Religions of God

There are several ways to understand the categories within which God is defined. Of course the first is the one defined as the category of theism. Here, God is usually but not always considered to be intervening, having the power to change things. Within this category are the monotheists and polytheists. Another may be considered the category of Deism in which God is either impartial (nonintervening), or perhaps even irrelevant to human concerns, and in some contexts with just about no interest in humanity whatsoever. Another category in which God is defined is Pantheism where God equals the universe and the Universe equals God. In Hinduism, there may be many deities, and in order to be in the presence of God one must be liberated from the cycle of birth. In Sikhism, individuals focus on a belief in the intimate relation with God, so that to have salvation God must be seen from the heart. In this sense the person's goal is revelation and communication or communion with God.

In Judaism God is Yahweh, and interacts with mankind in a personal way. In Christianity, God exists in three personas—the Father, the Son, and the Holy Spirit. In Islam, God is Allah, and is aware of everything in the universe including each person's private thoughts. Buddhism is nontheistic and its aim is to bring people out of suffering and into freedom and liberation so that ultimately people can become happy heavenly beings in a cycle of reincarnation.

The issue of whether a God does in fact exist is for theists, an issue as important as the phenomenological issue of life and death itself. For atheists, agnostics, and more or less unrepentant nonbelievers, the issue of whether there is an actual God does not reach the level of similar belief-significance as it does for theists, except in the atheistic-like reflex of offering pointed rejoinders about such a God's ostensible existence.

With certain categories of theists, as for example, sports minded people, the importance of God's presence is not only the belief, as for example that athletes show by gesticulating reverence both before and after they do something important in the game, and also in the sincerity with which sports-minded people generally treat such interest.

This was discussed by William Baker (2007), in his book *Playing with God: Religion and Modern Sport,* where Baker quotes Gary Smith in Smith's Sports Illustrated article entitled: *Blood Relations.* Smith (2006) reports a discussion between the brothers Max and Sam Kellerman regarding the origins of athletic spectacles. Max Kellerman, a TV/radio sports analyst and TV popular culture and news commentator, and Sam Kellerman, a writer/actor, were tuning in on this God issue. Baker, quoting Smith, indicates that Sam declared sport to be man's joke on

God. Sam said, "You see, God says to man, 'I've created a universe where it seems like everything matters, where you'll have to grapple with life and death and in the end you'll die anyway, and it won't really matter.'" So man says to God, "Oh yeah? Within your universe we're going to create a sub-universe called sports, one that absolutely doesn't matter, and we'll follow everything that happens in it as if it were a matter of life and death."

Whatever one can say about this brotherly conversation, it is God that sets the parameters of the discussion ("I've created a universe...."), and again, it is God who determines meaning ("....and it won't really matter."). But to continue the argument, man doesn't permit it to end there because man can also create a universe or at least a sub-universe which for some and in every way is an also strong universe and to boot, where in fact, everything does matter. And so, the thinking brain is the one that can imagine God's universe that matters greatly to some and not at all to others, and this thinking brain, this mind, also can create its own universe that perhaps might not at all matter, and at the same time yet completely matter.

It also should be pointed out that all people are psychologically uncomfortable with ambiguity and seek ways to untangle or resolve it. For most believers, the belief itself usually helps enormously in managing and even dissolving the tension associated with any ambiguity. It would seem that nonbelievers need to work harder for such resolution of tension.

Anatomical Tail to Thinking Brain

All in all, as can be seen, God is apprehended in different ways, and by different categories of people, and the function of God as related to the possibility of intervening in life is believed to exist in many different conceptualizations among varieties of peoples and cultures.

One thing is more clear however, and actually cannot be disputed—that without the thinking brain, God becomes only a being or an idea in the form of, as some definitions have it, a "nothing"—solely in virtual state, either extinct or extant. In virtual state, the crystallization of God, or the idea of God, or the "nothing," only awaits a thinking brain. Because of evolution this thinking brain becomes the gradually and ultimately emerging, transpositional successor, or even inheritor with respect to the function of the thinking brain's predecessor—the vestigial anatomical tail. As stated earlier, in this sense, a homologous relationship along the evolutionary trajectory is utilized here to trace function of organs rather than homologous parallel structures.

Then, God, or the idea of God that was conjecturally held in virtual state is either discovered as a true presence, or projected as solely the presence of God by the needful survival instinct created within the psyche of this thinking brain—a brain that is entirely epigenetically influenced—a psychoevolutionary phenomenon.

Chapter 7
The Inconclusion

So, for atheists and fellow travelers, the tail of animals that has become vestigial in evolution is then also functionally seen with respect to group behavior starting in lower forms on the phylogenetic scale all the way to higher-order primates and then into the evolutionary human psyche—a product of the cerebral cortex, the advanced thinking brain. Then, because of the existence of this thinking brain, this transmogrification is again further translated into the God in whose belief people are ostensibly enabled to gain all of the elements of peace of mind: psychological security, physical safety, emotional balance, personal empowerment, and the achievement of a world characterized by fairness. The entire transmogrified process is now both a foregone "tail," and perhaps a tale of God, with the possible alternate and tailor-made title for this book

The tale of God the tail, or, the tale of the tail

And this is not at all meant to cast aspersions on believers, or on a belief in God. In contrast, for believers in God, the basic thesis of this book connecting the vestigial tail in evolution with the presence of a thinking brain that actually locates God (either in the sky, in heaven, or in one's heart) cannot be in any sense an absolute refutation of the presence of an actual God because a case can be made that considers the formation of evolution as one that God created in the proverbial first place, that then evolves to eventually reach the thinking human so that this thinking mind can then find this self-same God that was there to create such evolution in the first place.

Even though because of the structure of the human brain, the origin of the concept of God may have been identified in this book as directly traced in evolution from an anatomical tail to the human ability to think and to conceptualize, nevertheless, as Stephen Jay Gould, the evolutionary scientist has remarked, that to be true to science at both polarities of considering whether an actual God exists, at one extreme is the absolute certainty of believers that indeed God does exist (which is scientifically speaking obviously based only on assumption), and at the other extreme are the agnostics, who at the very most can only say that they can't

H. Kellerman, *The Discovery of God: A Psychoevolutionary Perspective*,
SpringerBriefs in Psychology, DOI 10.1007/978-1-4614-4364-3_7, © The Author 2013

absolutely be sure either way. From a scientific standpoint, the atheist position is not one that can verify absence of a God but only state the belief of such an absence.

In any event, for most people, what we feel is usually far more compelling than what we think, and thus even the thinking brain frequently defers to the ascendancy of emotion; that is to say, the thinking brain defers to the brain in the gut. Therefore, inquiry, discussion, argument, and contemplation about the existence of a God, will, of course, continue. Thus, perhaps another alternate title for this book could be as follows.

> The discovery or origin of God only from a psychoevolutionary perspective, but not from other perspectives

Nevertheless, the extent to which the scientific argument is compelling is also framed only as a proposition; that is, it is necessary to identify those factors that constitute the most powerful theory to explain any phenomenon—in this case the phenomenon believed probably by the vast majority of people on Earth as the belief in God's existence.

The Most Powerful Theory

There is a widely accepted scientific basis for determining which theory among a variety of theories is the most powerful. This scientific basis or criterion for determining the most powerful theory concerns the absolute number of variables it takes to account for the widest array of phenomena so that the theory with the fewest number of variables accounting for the widest array of phenomena is scientifically considered the most powerful theory. Despite the fact that Newtonian theory is compelling, valuable, and powerful, Einsteinian theory is considered even more powerful because in Einsteinian theory, even fewer variables can explain an even wider array of phenomena. This is the idea of parsimony—Occam's razor—the law of economy or succinctness.

Thus, the proposition is asserted that the tail disappearing in evolution along with the eventual development of the cerebral cortex (the thinking brain) is what leads ultimately and perhaps even inexorably to the organism's conscious and complex concern with survival tensions of security and safety, as universal and ubiquitous existential concerns. These existential concerns motivate the thinking brain to seek a solution to this, what might be termed, ontological anxiety. Such survival concerns during the evolutionary process can be seen at all phylogenetic levels. For example, survival adaptations in evolution such as the function of the tail, or for that matter group ritual, convey survival and adaptational information facilitating in the example of group function, cohesion of the group. This cohesion includes the crystallization of roles for individual members. Interestingly, and even arguably, perhaps the main cohesive factors of civilized society include family influence, money, police, culture, and religion, although not necessarily in that order.

Finally, this evolution (so far, in contemporary life) develops into the appearance of the human thinking brain, and from there (as far as believers are concerned) to God (into the sky or the heart) where from there, and for many believers, God then presumably looks after us.

Therefore, here we essentially have two variables: the tail in lower phylogenetic levels, putatively transmogrified at higher levels to the human brain that can either project from its psyche or directly identify (find) the existing God, and thereby insist that the human wish for overall survival and therefore peace of mind is gained through God's protection.

Freudians who argue that the projection mechanism of the psyche is the basis for a belief in God, as well as a variety of atheists claiming the same thing, may be said to have a very powerful theory. But theists can also say that their theory (or certainty) is even more powerful. They can say it on the basis of the same scientific ground for identifying the most powerful theory: one variable—God—accounts for all phenomena.

However, atheists answer that sometimes an idea can be defined in so general a way and be so overarching that it can lose its meaning. Theists answer that their definition of God is not overarching, and its meaning is not infinitely general, and so it does not lose its meaning. Further, that another consideration in this assessment of which theory is most powerful is that even if one hits upon a fabulously powerful theory that because of its inherent power claims, at least, correlational correctness (powerful = correct), nevertheless, correlation is never actually equivalent or even unfailingly predictive of causation, and in some cases not at all predictive of causation.

Therefore, such doubt with respect to correlation and its possible association with causation can also apply to the atheist's position regarding the *function* of the tail in evolution and such *function's* proposed final resting place—as the brain. The brain as the tail's replacement (with respect to the tail as the former eye behind the head and now with the brain as the eye in the sky) makes such a theory seem quite powerful. But of course, that in itself doesn't necessarily make the theory valid. And this is similarly the same for the theist's position; that is, that just because one variable lays claim to be able to account for everything, this is also not necessarily correct.

On the other hand, there are instances where the correlational relationship becomes irresistible to refute. Such correlational consistency has such a seemingly intrinsic connection, that for all intents and purposes such correlation can, in fact, coincidentally constitute causation. Of course, "....can, in fact constitute causation," does not mean, "....does, in fact, constitute causation." And further, this claim of a highly organized and meaningful correlational matrix as it underscores the validity of a particular theoretical position or belief system as essentially equivalent to causation can be claimed both by theists and by atheists.

Ironically, the theist becomes the scientist in declaring that with one variable (God) all phenomena are accounted for, while the atheist, as scientist, can only claim an agnostic position, that of not knowing, and can scientifically go no further.

And then to jump to the atheistic claim that no God exists becomes a clear leap of faith! It is a faith that wears a mask of certainty of identifying God as inexistent and is a position, for example, favored by Richard Dawkins (2009), in his book, *The Greatest Show on Earth: The Evidence for Evolution.* And even though the atheist/ scientist will bet that on the basis of such strong evolutionary correlational evidence (and supposition), implying the certainty of the nonexistent or inexistent God—at that point of asserting such a position—the conclusion drawn, is more or less, a method, a way of thinking that is exactly logically congruent with the same method and way of thinking that the religionist has about the certainty of God's existence.

Hence, another alternate title for this book could be as follows.

The atheist's/scientist's faith in the tale of the tail
(The operative term being: "Faith")

Steven Pinker (2002), the atheist writer, cognitive scientist, and evolutionary psychologist, believes in universal moral imperatives that exist as logical products of the rational world—even the universe. In such a rational world, the dynamics are characterized by relationships between people. Timothy Keller (2008), who wrote *The Reason for God,* also values relationship and sees such relationship as even more highly valued with respect to redemption. In both cases, Pinker, and Keller, and whether atheist or theist, it seems to be that the value of relationship between people and how it is worked out and respected, becomes an agreed upon essence.

This thesis of the homologous transmutational phenomenon of *functionality* in evolution (seeking safety in attempting to ensure survival) is an example of the process examined in this volume that putatively traces anatomical evolutionary development (the vestigial tail to the thinking brain) here proposed as

The discovery of God: a psychoevolutionary perspective

Bibliography

Alcock, J. (2005). *Animal behavior: An evolutionary approach* (8th ed.). Sunderland, MA: Sinauer.

Angier, N. (2011, July 5). Thirst for fairness may have helped us survive. *New York Times: Science Times.*

Arlow, J. A. (1961). Ego psychology and the study of mythology. *Journal of the American Psychoanalytic Association, 9,* 371–393.

Armstrong, K. (2009). *The case for God.* New York, NY: Alfred A. Knopf.

Aron, L. (1996). *A meeting of minds.* Hillsdale, NJ: Analytic Press.

Aron, L. (2004). God's influence on my psychoanalytic vision and values. *Psychoanalytic Psychology, 21*(3), 442–451.

Atran, S. (2002). *In Gods we trust: The evolutionary landscape of religion.* New York, NY: Oxford.

Back, K. W. (1951). Social influences on persistence and change of attitudes. *Journal of Abnormal and Social Psychology, XLVI,* 9–23.

Baker, W. J. (2007). *Playing with God: Religion and modern sport.* Cambridge, MA: Harvard University Press.

Barrett, J. (2004). *Why would anyone believe in God?* Lanham, MD: Altamira Press.

Baylor Religion Survey (Fall, 2007). *Institute for studies of religion.* Baylor University.

Benedikt, M. (2007). *God is the good we do: A theology of Theopraxis.* New York, NY: Bottino Books.

Benjamin, S. P., & Zschokke, S. (2004). Homology, behavior and spider webs: Web construction behavior of *Linyphia hortensis* and *L. Triangularis* and its evolutionary significance. *Journal of Evolutionary Biology, 17,* 120–130.

Bion, W. R. (1959). *Experiences in groups.* New York, NY: Basic Books.

Blechner, M. (1988). Differentiating empathy from therapeutic action. *Contemporary Psychoanalysis, 24,* 301–310.

Bloom, P. (2004). *Descarte's baby.* New York, NY: Basic Books.

Bloom, P. (2005). Is God an accident? *The Atlantic Monthly,* 106–112.

Boyer, P. (2001). *Religion explained.* New York, NY: Basic Books.

Bracha, H. S. (2004). Freeze, flight, fight, faint: Adaptationist perspectives on the acute stress response spectrum. *CNS Spectrums, 9,* 679–685.

Buss, D. M. (2008). *Evolutionary psychology: The new science of the mind.* Boston, MA: Pearson.

Cartwright, D. (1968). The nature of group cohesiveness. In D. Cartwright & A. Zander (Eds.), *Group dynamics: Research and theory.* New York, NY: Harper and Row.

Cartwright, D., & Zander, A. (Eds.). (1968). *Group dynamics: Research and theory.* New York, NY: Harper and Row.

H. Kellerman, *The Discovery of God: A Psychoevolutionary Perspective,*
SpringerBriefs in Psychology, DOI 10.1007/978-1-4614-4364-3, © The Author 2013

Christakis, N. A., & Fowler, J. H. (2009). *Connected*. New York, NY: Little Brown & Co.

Confer, J. C., Easton, J. A., Fleischman, D. S., Goetz, C. D., Lewis, M. G., Perilloux, C., et al. (2010). Evolutionary psychology. *American Psychologist,* 110–126.

Dawkins, R. (2006). *The God delusion*. New York, NY: Houghton Mifflin.

Dawkins, R. (2009). *The greatest show on earth: The evidence for evolution*. New York, NY: The Free Press.

De Mello Franco, O. (1998). Religious experience and psychoanalysis: From man-as-god to man-with-god. *International Journal of Psychoanalysis, 79,* 113–131.

Dennett, D. C. (2007). *Breaking the spell: Religion as a natural phenomenon*. New York, NY: Viking/Penguin Group.

Eagleton, T. (2000). *The idea of culture*. Hoboken, NJ: Blackwell Publishing.

Eagleton, T. (2009). *Reason, faith, and revolution: Reflections on the God debate*. New Haven, CT: Yale University Press.

Festinger, L. (1950). Informal social communication. *Psychological Review, 57,* 271–282.

Feuerbach, L. (1941/1957). *The essence of Christianity*. New York, NY: Harper.

Flaherty, A. W. (2004). *The midnight disease*. New York, NY: Houghton Mifflin.

Foulds, G. A., Caine, T. Ma., & Crasey, M. A. (1960). Aspects of extra and intro-punitive expression in mental illness. *British Journal of Psychiatry, 106,* 599–610.

Fraley, R. C., Brumbaugh, C. C., & Marks, M. J. (2005). The evolution and function of adult attachment: A comparative and phylogenetic analysis. *Journal of Personality and Social Psychology, 89,* 731–746.

Frankl, V. (1969). *The will to meaning: Foundations and applications of Logo Therapy*. New York, NY: Penguin Books.

Freud, S. (1901/1914). *The psychopathology of everyday life*. London: T. Fisher Unwin.

Freud, S. (1907/1959). Obsessive actions and religious practices. In J. Strachey (Ed. and Trans.), *The standard edition of the complete psychological works of Sigmund Freud* (Vol. 9, pp. 115–127). London: Hogarth Press.

Freud, S. (1927/1961). The future of an illusion. In J. Strachey (Ed. and Trans.), *The standard edition of the complete psychological works of Sigmund Freud* (Vol. 21, pp. 1–56). London: Hogarth Press.

Garcia, C. L. (2007). Cognitive modularity, biological modularity and evolvability. *Biological Theory: Integrating Evolution, Development and Cognition, 2*(1), 62–73.

Glover, E. (1953). Team research on delinquency: A psychoanalytical commentary. *British Journal of Delinquency, 14*(3), 173.

Gould, S. J. (1991). Exaptation: A crucial tool for an evolutionary psychology. *Journal of Social Issues, 47,* 43–65.

Gould, S. J. (1997). The exaptive excellence of spandrels as a term and prototype. *Proceedings of the National Academy of Sciences of the United States of America, 94,* 10750–10755.

Gould, S. J., & Lewontin, R. (1979). The spandrels of San Marco and the Panglossian Paradigm Programme. *Proceedings of the Royal Society of London B, 205,* 581–598.

Graham, L. M. (1975). *Deceptions and myths of the bible*. New York, NY: Citadel.

Grossman, C. L. (2007). Faith and reason: A conversation about religion, spirituality and ethics. *USA Today*.

Grotstein, J. S. (2000). *Who is the dreamer and who dreams the dream? A study of psychic presences*. Hillsdale, NJ: Analytic Press.

Harris, S. (2004). *The end of faith*. New York, NY: W.W. Norton & Co., Inc.

Harris, S. (2006). *Letter to a Christian nation*. New York, NY: Knopf.

Haught, J. F. (2008). *God and the new atheism: A critical response to Dawkins, Harris, and Hitchens*. Louisville, KY: John Knox Press.

Hazelton, M. G., & Nettle, D. (2006). The paranoid optimist: An integrative evolutionary model of cognitive biases. *Personality and Social Psychological Review, 10,* 47–66.

Henig, R. M. (2007, March 4). Darwin's God. *The New York Times Magazine, Section 6,* 36.

Henrich, J., & Gil-White, F. (2001). The evolution of prestige: Freely conferred deference as a mechanism for enhancing the benefits of cultural transmission. *Evolution and Human Behavior, 22,* 165–196.

Hitchens, C. (1987) *Imperial spoils*: The curious case of the Elgin Marbles. Chatto and Windus, U.K.

Hitchens, C. (2007). *God is not great: How religion poisons everything*. New York, NY: Twelve Hachette Book Group.

Jha, A. (2010). Chimpanzees expand their territory by attacking and killing neighbors. *The Guardian, United Kingdom*, 12–14.

Johnston, M. (2009). *Saving God: Religion after idolatry*. Princeton, NJ: Princeton University Press.

Jones, E. (Ed.) (1974). The God complex. In *Psycho-myth, psycho-history: Essays in applied psychoanalysis* (Vol. 2, pp. 244–265). New York: Hillstone.

Josephs, L. (1988). A comparison of archeological and empathic modes of listening. *Contemporary Psychoanalysis, 24*, 281–300.

Kagan, J. (1996, September). Three pleasing ideas. *American Psychologist, 51*(9), 901–908.

Keller, T. (2008). *The reason for God*. New York, NY: Dutton, Penguin Group.

Kellerman, H. (1979). *Group psychotherapy and personality: Intersecting structures*. New York, NY: Grune & Stratton.

Kellerman, H. (1981). The deep structure of group cohesion. In H. Kellerman (Ed.), *Group cohesion: Theoretical and clinical perspectives*. New York, NY: Grune & Stratton.

Kellerman, H. (2008). *The psychoanalysis of symptoms*. New York, NY: Springer Science.

Kellerman, H. (2009a). *Dictionary of psychopathology*. New York, NY: Columbia University Press.

Kellerman, H. (2009b). *Love is not enough: What it takes to make it work*. Santa Barbara, CA: Praeger.

Kennedy, L. (2000). *All in the mind: A farewell to God*. London: Hodder & Stoughton, Ltd.

Kohut, H. (1984). *How does psychoanalysis cure?* Chicago, IL: University of Chicago Press.

Langer, S. K. (1967). *An introduction to symbolic logic* (3rd ed.) New York, NY: Dover Publications.

Lee, P. L. M., Clayton, D. H., Griffiths, R., & Page, R. D. M. (1996). Does behavior reflect phylogeny in swiftlets. *Proceedings of the National Academy of Sciences of the United States of America, 93*, 7091–7096.

Levenson, E. (1972). *The fallacy of understanding*. New York, NY: Basic Books.

Lifton, R. F. (1977). *Thought reform and the psychology of totalism: A study of brainwashing in China*. New York: W.W. Norton.

Lifton, R. J. (1979). *The broken connection*. New York, NY: Simon & Schuster.

Lifton, R. J. (1981). Historical and symbolic elements of social cohesion. In H. Kellerman (Ed.), *Group cohesion: Theoretical and clinical perspectives*. New York, NY: Grune & Stratton.

Lott, A. J., & Lott, B. E. (1965). Group cohesiveness as interpersonal attraction: A review of relationships with antecedent and consequent variables. *Psychological Bulletin, 64*, 253–309.

Love, A. (2007). Functional homology and homology of function: Biological concepts and philosophical consequences. *Biology and Philosophy, 22*(5), 691–708.

Mackie, J. L. (1982). *The miracle of theism: Arguments for and against the existence of God*. New York, NY: Oxford University Press.

Martyn, D. (2007). *Beyond deserving*. Grand Rapids, Michigan/Cambridge, UK: Wm. B. Eerdsmans Publishing.

Martyn, J. L. (1979). *History and theology in the fourth gospel, rev. and enl.*. Nashville, TN: Abingdon Press.

May, R. (1950). *The meaning of anxiety*. New York, NY: The Ronald Press Co.

May, R. (1983). *The discovery of being: Writings in existential psychology*. New York, NY: W.W. Norton & Co., Inc.

McCabe, H. (1987). *God matters*. London: Continuum.

McCarthy, K. (1990). Gallup Mirror of America Survey.

McIntosh, M. A. (1998). *A mystical theology: The integrity of spirituality and theology*. Malden, MA: Blackwell Publishing Ltd.

Meacham, J. (2009, April 13). The end of Christian America. *Newsweek Magazine*.

Meissner, S. J. (2009). The God question in psychoanalysis. *Psychoanalytic Psychology, 26*(2), 210–233.

Mitchell, S. (1988). *Relational concepts in psychoanalysis.* Cambridge: Harvard University Press.

Morse, C. (1994). *Not every spirit: A dogmatics of Christian belief.* Valley Forge, PA: Trinity Press International.

Muller, G. (2003). Homology: The evolution of morphological organization. In G. B. Muller & S. A. Newman (Eds.), *The organization of organismal form: Beyond the gene in development and evolutionary biology* (pp. 51–69). Cambridge: The MIT Press.

Murray, H. A., Erikson, E., & White, R. (1938). *Explorations in personality.* New York, NY: Oxford University Press.

Norenzayan, A. (2006). Evolution and transmitted culture. *Psychological Inquiry, 17*, 123–128.

Overman, D. (2009). *A case for the existence of God.* Lanham, MD: Rowman & Littlefield Publishers.

Pinker, S. (2002). *The blank slate: The modern denial of human nature.* New York, NY: Penguin.

Plutchik, R. (1980). *Emotion: A psychoevolutionary synthesis.* New York, NY: Harper & Row.

Plutchik, R. (2001, July–August). The nature of emotion. *American Scientist, 89*(4), 344–350.

Price, M. (2009). *DNA isn't the whole story. Monitor on psychology.* Washington, DC: American Psychological Association.

Ramachandran, V. S., & Blakeslee, S. (1998). *Phantoms in the brain* (p. 18). New York, NY: Morrow.

Redl, F. (1942). Group emotion and leadership. *Psychiatry, 5*, 573–596.

Reimarus, H. S. (1778). On the principal truths of natural religion. Charleston, SC: BiblioBazaar.

Rizzuto, A.-M. (1979). *The birth of the living God.* Chicago, IL: University of Chicago Press.

Rizzuto, A.-M. (1996). Psychoanalytic treatment and the religious person. In E. F. Shafranske (Ed.), *Religion and the clinical practice of psychology* (pp. 409–431). Washington, DC: American Psychological Association.

Rizzuto, A.-M. (1998). *Why did Freud reject God? A psychodynamic interpretation.* New Haven, CT: Yale University Press.

Rogers, C. R. (1951). *Client centered therapy: Its current practice, implications and theory.* Boston, MA: Houghton.

Rutledge, F. (2000). *Help my unbelief.* Grand Rapids, MI: Wm B. Eerdmans Publishing Co.

Sarna, N. (1986). *Exploring Exodus: The heritage of biblical Israel.* New York, NY: Schocken.

Schroeder, G. L. (2009). *God according to God: A physicist proves we have been wrong about God all along.* New York, NY: Harper Collins.

Scott, W. A. (1965). *Values and organization.* Chicago, IL: Rand McNally.

Scott, J. P. (1980). The function of emotions in behavioral systems: A systems theory analysis. In R. Plutchik & H. Kellerman (Eds.), *Emotion: Theory, research, and experience. The measurement of emotion* (Vol. 4, pp. 1–35). San Diego, CA: Academic Press.

Shaffer, J. B. P., & Galinsky, M. D. (1974). *Models of group therapy and sensitivity.* Englewood Cliffs, NJ: Prentice Hall.

Shermer, M. (1997). *Why people believe weird things: Pseudoscience, superstition, and other confusions of our time.* New York, NY: Henry Holt & Co.

Silver, M. (2008). The case against God. *New York: Jewish Currents Magazine, 40–44.*

Silverman, H. (2006 Autumn/2007 Winter). Positive atheism. *Humanistic Judaism, 28–29.*

Singer, R. D., & Fesbach, S. (1959). Some relationship between manifest anxiety, authoritarian tendencies, and modes of reaction to frustration. *The Journal of Abnormal and Social Psychology, 59*(3), 404–408.

Smith, G. (2006, April 17). Blood relations. *Sports Illustrated, 54–62.*

Sosis, R., & Alcorta, C. (2003). Signaling, solidarity, and the sacred: The evolution of religious behavior. *Evolutionary Anthropology, 12*(6), 264–274.

Spero, M. H. (1992). *Religious objects as psychological structures: A critical integration of objects relations theory, psychotherapy, and Judaism.* Chicago, IL: University of Chicago Press.

Stenger, V. (2007). *God, the failed hypothesis: How science shows that God does not exist.* Amherst, NY: Prometheus Books.

Stiedter, G. F., & Northcutt, R. G. (1991). Biological hierarchies and the concept of homology. *Brain, Behavior and Evolution, 38,* 177–189.

Strauss, D. F. (1835). *The life of Jesus critically examined.* 4th Edition. London: Swan Sonnenschein & Co.

Tobach, E., & Schneirla, T. (1968). The biopsychology of social behavior in animals. In Cooke, R. E. (Ed.), *The biologic basis of pediatric practice.* New York: McGraw Hill.

Thompson, C. (2009, September 27). Is happiness catching? *The New York Times Magazine.*

Tillich, P. (1963). *Christianity and the encounter of the world religions.* New York, NY: Columbia University Press.

Wade, N. (2009a). *The faith instinct: How religion evolved and why it endures.* New York, NY: Penguin Press.

Wade, N. (2009b, November 15). The evolution of the God gene. *The New York Times.*

Wagner, G. (2007). The developmental genetics of homology. *Nature Reviews: Genetics, 8,* 473–479.

Whitehead, A. F. (1979). *Process and reality: An essay in cosmology.* New York, NY: Free Press.

Wilson, E. O. (2012). *The social conquest of Earth.* New York, NY: Liveright Publishing Group, Inc./W.W. Norton.

Winnicott, D. W. (1953). Transitional objects and transitional phenomena: A study of the 1st Not Me Possession. *International Journal of Psycho-Analysis, 34,* 89–97.

Winnicott, D. W. (1965). The maturational process and the facilitating environment: Studies in the theory of emotional development. *The International Psycho-Analytical Library,* 64:1–276. London: The Hogarth Press.

Wood, J. (2009, August 31). God in the squad. *New Yorker Magazine,* 75–79.

Woodward, K. (1992, January 6). Why America prays. *Newsweek Magazine,* 197–202.

Wright, R. (2009a). *The evolution of God.* New York, NY: Little Brown.

Wright, R. (2009b, August 23). A grand bargain over evolution. *New York Times Op Ed.*

Wyers, E. J., Adler, H. E., Carpen, K., Chizar, D., Demarest, J., Flanagan, O. J., Jr., et al. (1980, November). The sociobiological challenge to psychology. *American Psychologist, 35,* 955–979.

Yalom, I. D. (1970). *The theory and practice of group psychotherapy.* New York: Basic Books.

Zeigarnik, B. V. (1927/1967). On finished and unfinished tasks. In W. D. Ellis (Ed.), *A sourcebook of Gestalt psychology.* New York, NY: Humanities Press.

Index

A

Adaptational theory, 14, 21, 25, 45, 54, 86
Affiliation, 23, 49, 50, 52, 55–56, 59, 61, 63
Agent detection, 22
Anger, 4, 17 18, 68,
Anxiety, 18–19, 24, 30, 43, 61, 79, 86 and God, 37
Assumption group, 59
Atheists, 12, 75, 77, 79, 87, 85–89
 ontological anxiety, 23–24
 philosophical arbitration, 81–83
 and theists, 23
 and fatalism, 23, 81

B

Belief
 adaptational theory, 21
 byproduct theory, 20
 defense mechanisms, 67
 faith-based belief, 50
 false, 69
 in God (*see* God)
 supernatural, 31, 68, 75
 thinking brain, 20
 transitional object, 73
 "unusual," 75
Bionian theory, 59
Blame psychology, 47, 49, 67
Brain, 11–17,19–22, 32, 42, 46, 49, 65, 83–87
Buddhism, 40, 83
Byproduct theory, 20

C

Causal reasoning, 22
Christians
 God existence, 83

Christologists, 43, 79
Classroom group, 53
 Cohesive group. *See also* Group
 affiliation psychology, 58
 belief system, 52
 co-opting pressure, 52
 core ideology, 53
 ego, relinquishment of, 54–55
 God-heads, 52
 God-idolatry, 52,
 "psyche" component, 52
 Superego alignment, 52
Cognitive tools,
 agent detection, 22
 causal reasoning, 22
 theory of mind, 22
Cohesion, 50,
 group, 51, 59
Correlation and causation,
 88–89, 75
Cultism, 58
Culture, 23, 36, 52, 54
 goals, 17
 group, 57, 59, 64
 supernatural, 75

D

Darwin, 45, 83
 survival, 24
Death, 18–19, 40, 67, 84
 Freud, 85
 God, 21
 instinct, 18
 ontological anxiety, 24
Defenses, 6
 compartmentalization, 31

H. Kellerman, *The Discovery of God: A Psychoevolutionary Perspective*,
SpringerBriefs in Psychology, DOI 10.1007/978-1-4614-4364-3, © The Author 2013